Methods and Tools in Biosciences and Medicine

Methods in Non-Aqueous Enzymology

Edited by

Munishwar Nath Gupta

Birkhäuser Verlag
Basel · Boston · Berlin

Editor

Prof. Dr. Munishwar Nath Gupta
Chemistry Department
Indian Institute of Technology, Delhi
Hauz Khas, New Delhi 110016
India

Library of Congress Cataloging-in-Publication Data
Methods in non-aqueous enzymology / edited by Munishwar Nath Gupta.
    p. cm. – (Methods and tools in biosciences and medicine)
    Includes bibliographical references and index.
    ISBN 3764358033 (alk. paper) – ISBN 0-8176-5803-3 (alk. paper)
      1. Enzymes. 2. Nonaqueous solvents. 3. Enzymology. 4. Lipase. I. Gupta, Munishwar
Nath. II. Series.

QP601 .M494 2000
572'.7–dc21   00-021710

Deutsche Bibliothek Cataloging-in-Publication Data
Methods in Non-Aqueous Enzymology / ed. by: Munishwar
Nath Gupta. – Basel ; Boston ; Berlin : Birkhäuser, 2000
    (Methods and tools, in biosciences and medicine)
    ISBN 3-7643-5803-3 (Basel...) Gb.
    ISBN 3-7643-6109-3 (Basel...) brosch.

ISBN 3-7643-5803-3 Birkhäuser Verlag, Basel – Boston – Berlin
ISBN 3-7643-6109-3 Birkhäuser Verlag, Basel – Boston – Berlin

Contents

List of Contributors

ANTHONSEN, THORLEIF, Department of Chemistry, Norwegian University of Science and Technology, N-7491 Trondheim, Norway; email: thorleif.anthonsen@chembio.ntnu.no

ARMISÉN, PILAR, Departimento de Biocatálisis, Instituto de Catalisis, CSIC, Campus Universidad Autónoma, E-28049 Madrid, Spain

BORNSCHEUER, UWE T., Institute for Chemistry and Biochemistry, Department of Technical Chemistry and Biotechnology, University of Greifswald, D-17487 Greifswald, Germany; e-mail: bornsche@mail.uni-greifswald.de

BOSLEY, JOHN A., Unilever Research Colworth, Sharnbrook, Bedford MK44 1LQ, UK; e-mail: john.bosley@unilever.com

CAMINAL, GLORIA, Department of Chemical Engineering, Facultat Ciències. Edifici C., Universitat Autónoma de Barcelona, E-08193 Bellaterra, Spain

CHOPINEAU, JOËL, Laboratoire de Technologie Enzymatique, ESA CNRS 6022, Université de Technologie de Compiègne, F-60206 Compiègne cedex, France; e-mail: joel.chopineau@utc.fr

CLAPÉS, PERE, Department of Peptide and Protein Chemistry, Chemical and Environmental Research Insitute-C.S.I.C., E-08034 Barcelona, Spain; e-mail: pcsqbp@cid.csic.es

DANIELSSON, BENGT, Pure and Applied Biochemistry, Center for Chemistry and Chemical Engineering, Lund University, S-22100 Lund, Sweden

DOMURADO, DOMINIQUE, Groupe de Pharmacocinétique des Prodrogues et Conjugués Macromoléculaires (INSERM), Centre de Recherche sur les Biopolymères Artificiels, URA CNRS 1465, Faculté de Pharmacie, 34060 Montpellier cedex 2, France

FELIU, JOSEP ANTÓN, Department of Chemical Engineering, Facultat Ciències. Edifici C., Universitat Autónoma de Barcelona, E-08193 Bellaterra, Spain

FERNÁNDEZ-LAFUENTE, ROBERTO, Departimento de Biocatálisis, Instituto de Catalisis, CSIC, Campus Universidad Autónoma, E-28049 Madrid, Spain

FERNÁNDEZ-LORENTE, GLORIA, Departimento de Biocatálisis, Instituto de Catalisis, CSIC, Campus Universidad Autónoma, E-28049 Madrid, Spain

GUISÁN, JOSÉ M., Departimento de Biocatálisis, Instituto de Catalisis, CSIC, Campus Universidad Autónoma, E-28049 Madrid, Spain

GUPTA, MUNISHWAR NATH, Department of Chemistry, Indian Institute of Technology, Delhi, New Delhi 110016, India; e-mail: mn_gupta@hotmail.com

KHARE, SUNIL KUMAR, Central Institute of Agricultural Engineering, Indian Council of Agricultural Research, Nabi-bagh, Bhopal-462038, India

LAGOUTTE, BERNARD, Service de Bioénergétique, URA 2096 CNRS, Département de Biologie Cellulaire et Moléculaire, Centre d'Etudes de Saclay, 91191 Gif-sur-Yvette cedex, France

LÓPEZ-SANTÍN, JOSEP, Department of Chemical Engineering, Facultat Ciències. Edifici C., Universitat Autònoma de Barcelona, E-08193 Bellaterra, Spain

MATEO, CESAR, Departimento de Biocatálisis, Instituto de Catalisis, CSIC, Campus Universidad Autónoma, E-28049 Madrid, Spain

NAKAJIMA, MITSUTOSHI, National Food Research Institute, Ministry of Agriculture, Forestry and Fisheries, Tsukuba, Ibaraki 305–8642, Japan; e-mail: mnaka@nfri.affrc.go.jp

OTTOLINA, GIANLUCA, Istituto di Biocatalisi e Riconoscimento Molecolare, C.N.R., I-20131 Milano, Italy; e-mail: ottolina@ico.mi.cnr.it

PEILOW, ALAN D., Unilever Research Colworth, Sharnbrook, Bedford MK44 1LQ, UK; e-mail: alan.peilow@unilever.com

RAMANATHAN, KUMARAN, Pure and Applied Biochemistry, Center for Chemistry and Chemical Engineering, Lund University, S-22100 Lund, Sweden; e-mail: kumaran.ramanathan@tbiokem.lth.se

REES JÖNSSON, BIRGITTA, Pure and Applied Biochemistry, Center for Chemistry and Chemical Engineering, Lund University, S-22100 Lund, Sweden

RIVA, SERGIO, Istituto di Biocatalisi e Riconoscimento Molecolare, C.N.R., I-20131 Milano, Italy; e-mail: rivas@ico.mi.cnr.it

RODA, GABRIELLA, Istituto di Biocatalisi e Riconoscimento Molecolare, C.N.R., I-20131 Milano, Italy

SABUQUILLO, PILAR, Departimento de Biocatálisis, Instituto de Catalisis, CSIC, Campus Universidad Autónoma, E-28049 Madrid, Spain

SADANA, AJIT, Chemical Engineering Department, University of Mississippi, University, MS 38677–9740, USA; e-mail: cmsadana@olemiss.edu

SJURSNES, BIRTE J., Østfold College, N-1702 Sarpsborg, Norway

SNAPE, JONATHAN B., Mylnefield Research Services, Scottish Crop Research Institute, Invergowrie, Dundee, DD2 5DA, Scotland, UK

THOMAS, DANIEL, Laboratoire de Technologie Enzymatique, ESA CNRS 6022, Université de Technologie de Compiègne, F-60206 Compiègne cedex, France

Preface

Extending the range of enzymatic catalysis by using non-aqueous media has now developed into a powerful approach in biochemistry and biotechnology. One peculiar feature which distinguishes it from the conventional enzymology (carried out in aqueous buffers) is that the awareness of different parameters that control and influence the behaviour of enzymes in such environments has emerged rather slowly. Science is about being able to repeat what somebody else has done. Absence of knowledge about such well-defined parameters/factors has sometimes made some workers rather cautious and diffident about using this approach in their laboratories. But for this, non-aqueous enzymology would be more widely practised. It is these thoughts that made me feel that the availability of some well-defined protocols for various applications involving enzymes in non-aqueous environments would further catalyze the growth of this area. Hence this book, in which each chapter has some protocols in a specific area. The protocols are preceded by brief background material. The early chapters, which are of general importance, concern control of water activity and stabilization via immobilization. Some subsequent chapters provide the protocols for transformations involving lipids and carbohydrates, peptide synthesis, and preparation of chiral compounds. The disproportionate focus on lipases is not a coincidence; this class of enzymes has been used more often than others in non-aqueous enzymology. Use of reverse micelles as microreactors and use of biosensors in harsher non-aqueous environments are two chapters that are futuristic in the sense that they probably represent the tip of an emerging iceberg. The first and last chapter do not contain protocols; both provide general perspectives. The first one to the whole area and the last one outlines the role which various factors in concert with the medium may play in protein folding in both *in vitro* and *in vivo* contexts.

I am pleased to have been able to entice some very good workers and distinguished scientists in writing the individual chapters. I am grateful that they consented to join me in this endeavour. One nice summer morning in Compiègne, France, Professor D. Thomas (one of the very early people to be approached) very graciously and promptly agreed to contribute a chapter. Within the next few days, thanks to the marvel of e-mail, most of the other chapters were also finalized. It was nice of Birkhäuser Verlag to agree so promptly to publish this book. I have been lucky in having kind, competent and patient help, first from Dr. Petra Gerlach and later on from Dr. Claus Puhlmann, both from Birkhäuser, Basel.

I should acknowledge the support of the Department of Science and Technology (Government of India) for funding my research work in this area. Also, the generous support of the Department of Biotechnology (Government of India) and Council of Scientific and Industrial Research (both extramural and technology mission wings) has made it possible for me to continue my work in applied enzymology in general. Without all this, I would not have thought of, or ventured into, this project.

A modest book-writing support grant from IIT, Delhi is also acknowledged. Ms Ipsita Roy, a graduate student in my lab, must be thanked for her considerable help with word processing and keeping me connected via e-mail etc. to various people involved with the book. Two other members of my research group, Dr. Sunita Teotia and Ms. Shweta Sharma were also of great help in the final stages of editing the various chapters. My family, my wife Dr. Sulbha Gupta and son Chetan Gupta, have patiently tolerated my devotion of 'extra time' to the book.

I hope it has been worth it. I hope that the book will be found useful by those who use it.

(M.N. GUPTA)
February 2000
New Delhi, India

1 Non-Aqueous Enzymology: Issues and Perspectives

Munishwar N. Gupta

Contents

1   Introduction

The practice of using enzymes in media other than pure aqueous buffers is not new. One can easily identify three distinct phases in the growth of this area. These phases along with the milestones are indicated in Table 1.

Table 1 Phases of enzyme use in media

Phase	Time period	Some significant milestones
Early work -sporadic attempts	1913	Bourquelot and Bridel [1] Stereospecific synthesis of various alkyl glucosides using dry emulsion in alcohol.
	1966–67	Dastoli and co-workers [2,3] Chymotrypsin and Xanthine oxidase powders in dry organic solvents.
Renaissance	1980–1990	The seminal work of Zaks and Klibanov [4,5]. The work emanating from the group of Adlercreutz and Mattiasson [6,7]. Medium Engineering rules framed [8]. The importance of thermodynamic water activity realized [9].
Non-aqueous enzymology as an established area in biotechnology	The Nineties	Option of enzyme-based synthesis establishes itself. Attempts to unravel enzyme structure in non-aqueous media by physicochemical methods start. Possibilities of using whole cells are explored.

Methods and Tools in Biosciences and Medicine
Methods in non-aqueous enzymology, ed. by M. N. Gupta
© 2000 Birkhäuser Verlag Basel/Switzerland

Today, enzymes can be used (in addition to the traditional way of using these in aqueous buffers) in water-organic co-solvent mixtures, nearly-anhydrous organic solvents, reverse micelles, supercritical fluids and organic solvent vapors. There have been efforts to optimize both their activities and stabilities under these different conditions. Table 2 shows the various forms in which enzymes are used today in non-aqueous environments.

Table 2   Various forms of enzymes being used in non-aqueous media

Form	Reference
Suspension in the powder form	Zaks and Klibanov [4,5]
Polyethylene glycol-linked enzymes (soluble)	Inada et al. [10] Bovara et al. [11] Ljungar et al. [12]
Entrapped in surface-modified polymeric granules	Khmelnitsky et al. [13]
CLEC™	Margolin [14], Lalonde et al. [15]
Enzyme-polymer noncovalent complexes (both soluble and insoluble forms)	Otamari et al. [16], Virto et al. [17]
Enzymes in supercritical fluids	Kamat et al. [18]
Lipid-coated enzymes	Okahata and Mori [19]
Enzyme-polyelectrolyte complexes	Kudryashova et al. [20]
Enzyme paired with hydrophobic ions	Kendrick et al. [21]

This table is meant to be illustrative rather than comprehensive. Other ways of using enzymes in non-aqueous media are described in later chapters of this book. The most important one is that of using enzymes in the immobilized form. It is not possible to suggest which form would be best for all kinds of situations or applications. The enzyme being used, the optimum range of water content for that particular application and the kind of solvent to be used, are some of the considerations that may automatically preclude some forms. Thus, while designing a particular protocol, care should be taken to consider all available options.

In the case of water-miscible organic solvents, the general situation still is that one can use organic solvent up to 30–40% in cases of relatively more stable enzymes/modified enzyme forms. In many cases, it is also possible to use the range in which organic solvent concentration is 70–99.9%. The middle range of concentration is not normally available, but at this moment, it does not look as if this is a serious limitation. Nevertheless, a greater understanding of inactivation mechanisms and protein-solvent interactions relating to this range would be welcomed.

In recent years, low water-systems have attracted considerable attention [22–25]. Several advantages associated with such systems have been discussed [13, 22, 26]. Enhanced thermostability of enzymes, medium engineering rules and various approaches used for increasing enzyme stability in anhydrous solvents have also been reviewed [6, 22, 26]. In this overview, the focus is on several aspects in which considerable development has taken place in the last few years. Individual applications and some other general aspects are described in the remaining chapters of the book.

2   Enzyme structure in non-aqueous environments

Considerable efforts have been made to probe the structure of enzymes at limited water concentrations by various physicochemical tools. Klibanov [24] summarized the results from FT-IR spectroscopy: "Placing lyophilized subtilisin in a variety of anhydrous solvents... has no appreciable effect on its secondary structure (as reflected by the α-helix content). However, in the preceding step, lyophilization results in a significant (and reversible, if lyophilization is done carefully) denaturation of subtilisin and many other proteins". However, if subtilisin is dissolved in non-aqueous solvents, the changes in α-helix or β-sheet structures do occur in some cases [the study was limited to only four solvents] and these correlate very well with the enzymatic activity [27]. Use of CD and amide proton exchange monitored by 2D NMR have shown that lysozyme dissolved in glycerol with 1% water has a highly-ordered native-like structure with the stability exceeding that in water [28].

Recently, Schmitke et al. [29] have looked at the X-ray crystal structures of subtilisin in 40% acetonitrile, 20% dioxane and neat organic solvents. The same group [30] has also compared the X-ray crystal structures of an acyl-enzyme intermediate of subtilisin formed in anhydrous acetonitrile and water. An important result reported by Partridge et al. [31], using proton solid-state NMR relaxation measurements with subtilisin suspended in cyclohexane, dichloromethane or acetonitrile, is that hydration is the primary determinant of structural mobility and medium dielectric may not play an important role. The structure was rigid up to a_w (water activity) equal to 0.22. Early work using NMR has been reviewed elsewhere [22].

In the coming years, no doubt we will have a greater understanding of structure-function correlations leading to less empirical designs of applications.

3   pH memory

In 1986, Klibanov observed that the "enzyme remembers the pH of the latest aqueous solution to which it has been exposed" [26]. This pH memory meant that the highest activity in anhydrous organic solvents occured when the enzyme had been previously lyophilized from a buffer at a pH corresponding to the optimum pH of the enzyme in aqueous systems. The beneficial effect of this pH-tuning could be fairly impressive, e.g. in the case of transesterification rates in subtilisin-catalyzed reactions, pH tuning led to a 300-fold increase [26]. Two early opinions regarding the origin of this phenomenon were: (1) The existence of kinetically trapped enzyme structure in organic solvents [5]; (2) that pH tuning determines the ionization state of the amino acid residues in the enzyme [32]. A curious facet of the phenomenon was the observation that the pH memory of chymotrypsin applies not only to its catalytic activity but also to thermal stability [5]. The general nature of this was indicated by the result – the half life of subtilisin BPN in dimethyl formamide dramatically depends on the pH of the aqueous solution from which the enzyme was lyophilized, increasing from 48 min to 20 hrs, when the pH was raised from 6.0 to 7.9 [33].

As far as the effect of "pH-tuning" on enzyme activity is concerned, the data in the literature indicated that the explanation offered by Mattiasson and Adlercreutz [32] is likely to be the correct one. It had been reported that the resonance frequency of ^{15}N-enriched solid preparations of free imidazole, histidine, or histidine in the active site of lytic protease, is dependent on the pH from which the amino acid or the enzyme was prepared [34, 35]. More recently Costantino et al. [36] using FT-IR confirmed that the ionized state of the side chains of amino acids in proteins persists after lyophilization.

Yang et al. [37] showed that the "apparent pH dependence" of the activity of subtilisin dispersed in organic solvents was also dependent upon solvent type, water content and method of water delivery. Around the same time [38] it was reported that the activity of bilirubin oxidase in an anhydrous organic solvent varied significantly with the amount of buffer and nature of buffer during the "tuning" process. The lyophilization from a volatile buffer gave an enzyme that showed no activity in the organic solvents. Blackwood et al. [39] showed that the buffers soluble in the organic solvents are able to erase pH memory and control the activity independent of earlier aqueous pH. The same group [40] later synthesized functionalized dendritic polybenzylethers as acid/base buffers for use in toluene and showed that these buffers could "alter and precisely control the ionization state of biocatalysis" in the cases of subtilisin and chymotrypsin.

The demystification of the phenomenon of pH memory is perhaps completed by the latest results from this group [41]. It has been shown that the "volatile buffers can override the pH memory of subtilisin in the organic media." Thus, it is essentially a question of the enzyme retaining the right charges on the residues involved in the catalysis.

However, Guinn et al. [42] had reported that for three pH values [2.0, 7.5 and 11.0), alcohol dehydrogenase showed highest activity in heptane when it was lyophilized from a solution of pH 2.0. The enzyme has little activity in water at this pH and its pH optimum in water is 7.5.

4   Molecular imprinting in anhydrous organic solvents

A concept related to pH memory is when an enzyme/protein remembers the exposure to a ligand. Russel and Klibanov [43] showed that "lyophilization of subtilisin from aqueous solution containing competitive inhibitors (followed by their removal) created an enzyme that was up to 100 times more active than the enzyme lyophilized in the absence of such ligands". A few years later, two different groups used this idea to develop bio-imprinting [44, 45] or molecular imprinting [46, 47]. Stahl et al. [44] showed that the precipitation of a complex of chymotrypsin and N-acetyl-D-tryptophan from aqueous solution by n-propanol created the enzyme that could esterify the D-amino acid at a significant rate. Braco et al. using bovine serum albumin [46] found that "the resultant imprinted protein preparation binds up to 30-fold more of the template compound in anhydrous solvents than the non-imprinted protein...". A later study [47] implicated "hydrogen bonds as the driving force" and showed that "many H-bond-forming-macromolecules other than proteins, such as dextrans and their derivatives, partially hydrolyzed starch and poly (methacrylic acid) could also be imprinted..." An increase in total water concentration by 4 mM [44], the presence of 0.4% water in the solvent [43], and the addition of 2% water [46] were reported to erase the memory in different systems. Prior to this Klibanov [48] had emphasized that induction of enzyme memory requires water wherein considerable conformational flexibility in the protein exists, whereas high conformational rigidity observed in nearly anhydrous environments is a prerequisite for display and retention of memory.

In the case of enzymes, imprinting can also be used to create memory for ligands that bind to sites other than active sites. The ligands include certain sugars [49] and salts [50]. Mingarro et al. [51] described interfacial activation-based molecular bio-imprinting to improve the efficiency of some lipases and phospholipase A_2 in non-aqueous environments.

Mishra et al. [52] used FT-IR to detect significant differences in the secondary structure of lysozyme, chymotrypsinogen and BSA upon imprinting. Griebenow and Klibanov [53] showed (again by using FT-IR) that the mechanism of subtilisin activation by KCl and N-acetyl-L-phenylalanine-amide may be due to preservation of the induced secondary structure in organic solvents.

Recent work by Stewart et al. [54] further explores the additional consequences of molecular imprinting. Chymotrypsin co-lyophilized with the competitive inhibitors, N-acetyl-L-tryptophan or N-acetyl-D-tryptophan was completely protected from inactivation when reacted with iodomethane *in vacuo*. However, imprinting with indole and sorbitol facilitated inactivation by iodomethane. In both cases, imprinting ligands were removed before reaction with iodomethane.

Ke and Klibanov [55] have reported that the reaction rates in non-aqueous solvents depend upon whether the enzyme was lyophilized or precipitated by organic solvents. "The magnitude of this dependence was markedly affected by the nature of the solvent and enzyme". Considering that the enzyme activity in non-aqueous media is shown to depend upon "enzyme history", the results are relevant to the phenomena of enzyme memory and molecular imprinting.

An avoidable confusion exists in the literature concerning the term "molecular imprinting". The term has also been used by Mosbach et al. in their extensive work on molecularly imprinted polymers. The technique in this case is very different and has been reviewed in a number of places e.g. by Mosbach and Ramstrom [56].

Mention should also be made of yet another bio-imprinting approach by Soler et al. [57]. It was shown that the refolding of urea-denatured chymotrypsin-agarose derivatives in 20% DMF, both in the presence and absence of the substrate, yielded enzyme derivatives with different kinetic parameters. An important factor was that the imprints were stable in aqueous buffers.

5   Using antibodies in organic solvents

Recognition of an antigen by the antibody molecule involves very similar interactions to those involved in binding a substrate to the enzyme. Russel and Klibanov [58] showed that antigen binding to the antibody takes place in organic solvents. Thereafter, Weetall [59] used immobilized antibodies in hexane. It was soon realized that such systems provide good opportunities to carry out immunoassays of water insoluble substances. Analysis of pesticides and their residues in environments constituted an obvious and useful application.

Stocklein et al. [60] looked at the binding of triazine herbicides to antibodies in anhydrous organic solvents. Matsuura et al. [61] evaluated the performance of 20 mouse monoclonal antibodies against okadaic acid in organic solvents. Okadaic acid occurs in some marine sponges and has been identified as one of the toxic hydrophobic compounds causing diarrhetic shellfish poisoning. The results were quite promising, with two monoclonals functioning well in the presence of some organic solvents. No relationship between the subclass of the immunoglobulin and the binding activity of the antibody in organic solvents was observed.

It should be added that antibody-like affinities and selectivities are becoming possible in the case of imprinted polymers. The results with beta-blocker S-propranolol showed that the radioligand-binding assays could be performed in toluene and exhibited excellent enantioselectivity [62, 63].

Such imprinted polymers also circumvent the stability problem of the antibodies in non-aqueous milieu. Important work by Janda et al. [64] addressed the issue of stability and investigated immobilization with appropriate coupling procedure as a means to enhance the stability of catalytic antibodies in organic solvents. They reported that "site-directed coupling through the carbohydrate functionality has proven to be quite disastrous, while heterobifunctional crosslinking has been reproducible". Using a succinimide coupling method and controlled pore glass beads as matrix, they successfully immobilized eleven catalytic antibodies against esters. While some enhanced stability in few co-solvents was observed, attempts to catalyze transesterification under anhydrous conditions were unsuccessful.

Wang et al. [65] described polymers prepared from aminoglucose-based monomers and conjugated some of these with an antibody against aldrin. They reported that the conjugate "was competent for 5 hrs in acetonitrile, methanol and 2-propanol with 96, 60 and 57% of the original binding, respectively, while the native antibody retained no binding ability under identical conditions". More recently, Okahata et al. [66] described the preparation of lipid-coated catalytic antibody that was soluble in organic solvents and "showed a remarkable reactivity of lipophilic esters in a buffer solution containing 20–80% dimethyl sulfoxide". The native antibody had very low k_{cat} value in the cosolvent mixture.

It looks certain that in coming years, the range of many immunochemical techniques will be extended through the use of non-aqueous media. Also, as forms of catalytic antibodies stable in the non-aqueous milieu become available in increasing numbers, many more biotransformations will become possible. The critical issue again in this context as well is that of stability of the protein molecule in non-aqueous environments.

6   Catalysis by whole cells in organic media

Some of the earlier work in this area is by Nikolova and Ward [67, 68] and Nakamura et al. [69]. A recent paper by Griffin et al. [70] provides a good discussion on the possibilities and advantages of using whole cells in organic media. The cells have to be either entrapped or immobilized on a solid support to reduce mass-transfer limitations since whole cells tend to form clumps in organic solvents. The advantages of using whole cells as such for catalysis include: (1) The saving of downstream processing costs for the enzymes; (2) Cofactor regeneration may be possible more easily, and (3) some enzyme activities in fact survive harsher conditions better when still within the cell. Griffin et al. [70] studied the asymmetric reduction of acetophenone with alginate-entrapped yeast in hexane. The alcohol dehdrogenase within yeast catalyzed the reduction. For regeneration of NADH, a regeneration scheme using 2-hexanol as a "sacrificial cosubstrate" was used to recycle the NAD(H). The values of initial rate and enantiomeric excess (ee) were optimum at pH 3 and 10. The minimum mass transfer limitation was observed with the alginate beads in the size range of 1.4–2 mm. The maximum initial rate required water content in the range of 3–8 g per g of dry cell weight. However, the ee had a value of 0.98–1 for all water contents tested.

Another interesting system has been described by Andersson et al. [71, 72]. A hydrogenase contained in *Alcaligenes eutrophus* was evaluated for regeneration of NADH with H_2 gas as the reduction source. The primary reactions studied were alcohol dehydrogenase catalyzed reduction of cyclohexanone and lactate dehydrogenase catalyzed reduction of pyruvate. EDTA/toluene-treated cells were either immobilized on celite or entrapped in alginate beads. For reduction of cyclohexanone, celite immobilized catalysts were tried in six different organic solvents (with 20% v/v added water) and the best product yield was obtained with butyl acetate. The authors were forthright enough to mention that NADH regeneration observed was more likely due to alcohol impurity in the solvent [71].

Subsequently, heptane, toluene and trichloroethane were found to be suitable for the coupled reaction catalyzed by alcohol dehydrogenase and the cells [72]. The optimization of cyclohexanol yield ($\geq$ 99%) was achieved by non-immobilized CTAB-permeabilized cells in heptane containing 10% water. The advantage of this approach over co-substrate-driven regeneration is that no by-products are left in the system.

Now that pioneering work is available in the literature that shows considerable promise, the initial diffidence is likely to be replaced with greater activity in this area.

7   Protein folding in non-aqueous media

Various aspects of protein folding including the importance of medium are discussed by Sadana in the last chapter of this book (Chapter 12). It is, however, worthwhile to mention here that intracellular conditions in extremophiles may involve water activities as low as 0.6. Also, in the case of halophiles, "the few proteins that have been investigated require salt for folding and assembly" [73]. Recently, Rariy and Klibanov [74] reported that refolding of hen egg white lysozyme to active structure is markedly enhanced by the presence of common salts.

It seems likely that as our understanding of protein-solvent interactions increases it will be possible to "medium-engineer" the protein refolding.

8   Ultrasonoenzymology in non-aqueous media

The range of ultrasound is generally considered to be between 20 kHz to beyond 100 MHz. Ultrasonochemistry is now a fairly established area wherby frequencies between 20 kHz and 40 kHz have been used to accelerate reactions. The basic principle involved is that ultrasound generates cavitation within a liquid and cavitation in turn provides energy to enhance reaction rates of chemical processes [75]. A very good article on the basic physics of cavitation by a well known name in this area has been published recently [76].

Prior to this, Sinisterra reviewed the applications of ultrasound in biotechnology [77]. He describes the earlier results concerning deactivation of papain and albumin when subjected to ultrasonic frequencies. It is therefore understandable that the use of ultrasound in enzyme-catalyzed reactions has been rather slow to take off. However, some valuable results have been obtained. Sinisterra [77] also reviewed the effect of the nature of the solvent on enzyme-catalyzed peptide synthesis. The inactivating effect of the hydrophic solvents is enhanced presumably due to greater collision frequency of the solvent contacting the enzyme as a result of ultrasonication. Also, an interesting observation is that "ultrasound could increase the degree of dissociation of the acid organic molecule affecting the pH of the microenvironment and thus changing the enzymatic activity."

Conversely, Vulfson et al. [78] report that "subtilisin was found to be much more resistant to inactivation by ultrasound irradiation in organic solvents than in water." They also found that continuous sonication during transesterification reaction led to a greater increase in reaction rate as compared to pretreatment of the enzyme with ultrasound. The increase in water content of the medium (alcohols in this case) potentiated the effect of ultrasonication. More recently, Lin and Liu [79] reported that rates of lipase-catalyzed reactions increased several-fold by using ultrasonics while retaining stereoselectivity of the process.

More data is needed before the ground rules for judicious use of ultrasonics for influencing enzyme catalysis in non-aqueous solvents can emerge.

Another approach that is likely to see greater application in this area is the use of microwaves to increase the rate of enzyme-catalyzed reactions. Parker et al. [80] reported that irradiation of a lipase suspension in organic media with microwaves (2.45 GHz) enhanced the reaction rate by 2- to 3-fold over the control.

9 Conclusions

Where do we go from here? This is one area where 'applications' are ahead of the understanding of basic principles. We are, however, catching up fast on the latter. There is enough information available on protein-ligand interactions in general but there is still inadequate knowledge about protein-solvent interactions. This is definitely a rate-limiting step in the further growth of this area.

A universal medium parameter that can be used to chose an organic solvent for optimum activity or predict the stability of a particular enzyme is still not evident. If this remains for low water systems, the situation is more dismal in the case of water-organic cosolvent mixtures. We have only just begun to evaluate how the presence of an "impurity" in the enzyme preparation may influence its stability and catalytic efficiency.

Temperature is one parameter that is bound to be studied more for making non-aqueous enzymology more versatile. Working with organic solvents makes cryoenzymology possible. The use of temperature to improve even selectivity in such media has begun to appear in the literature [81, 82].

All in all, it has been an exciting decade for non-aqueous enzymology. With greater appreciation of basic biochemical and biophysical principles governing enzyme behaviour in such media, the next decade is bound to be even more exciting.

Acknowledgments

I thank Dr. P. Adlercreutz [Sweden] for generously sharing with me recent results from his laboratory in the form of various theses and reprints. This was helpful in preparation of this manuscript.

References

1   Khmelnitsky YL, Levashov AV, Klyachko NL, Martinek K (1988) Engineering biocatalytic systems in organic media with low water content. *Enzyme Microb Technol* 10: 710–724
2   Dastoli FR, Musto WA, Price S (1966) Reactivity of active sites of chymotrypsin suspended in an organic medium. *Arch Biochem Biophys* 115: 44–47
3   Dastoli FR, Price S (1967) Catalysis by xanthine Oxidase suspended in organic media. *Arch Biochem Biophys* 118: 163–165
4   Zaks A, Klibanov AM (1988) Enzymatic catalysis in non-aqueous solvents. *J Biol Chem* 363: 3194–3201
5   Zaks A, Klibanov AM (1988) The effect of water on enzyme action in organic media *J Biol Chem* 263: 8017–8021
6   Adlercreutz P, Mattiasson B (1987) Aspects of biocatalyst stability in organic solvents. *Biocatalysis* 1: 99–108
7   Reslow M, Adlercreutz P, Mattiasson B (1988) The influence of water on protease-catalyzed peptide synthesis in acetonitrile/water mixtures. *Eur J Biochem* 172: 573–578

8   Laane C (1987) Medium engineering for bioorganic synthesis. *Biocatalysis* 1: 17–32
9   Halling PJ (1987) Rates of enzymatic reactions in predominantly organic, low water systems. *Biocatalysis* 1: 109–115
10 Inada Y, Takahashi K, Yoshimoto T et al. (1986) Application of polyethlene glycol-modified enzymes in biotechnological processes: organic solvent-soluble enzymes. *Trends Biotechnol* 4: 190–194
11 Bovara R, Carrea G, Gioacchini AM et al. (1997) Activity, stability and conformation of methoxypoly (ethylene glycol) -subtilisin at different concentrations of water in dioxane. *Biotechnol Bioeng* 54: 50–57
12 Ljungar G, Adlercreutz P, Mattiasson B (1993) Reactions catalyzed by PEG-modified α-chymotrypsin in organic solvents. Influence of water content and degree of modification. *Biocatalysis* 7: 279–288
13 Khmelnitsky YL, Neverova IN, Gedvovich AV et al., (1992) Catalysis by α-chymotrypsin entrapped into surface-modified polymeric granules in organic solvents *Eur J Biochem* 210: 751–757

14 Margolin AL (1996) Novel crystalline catalysts. *Trends Biotechnol* 14: 223–230

15 Lalonde JJ, Navia MA, Margolin AL (1997) Cross-linked enzyme crystals of lipases as catalysts for kinetic resolution of acids and alcohols. *Methods Enzymol* 286: 443–464

16 Otamiri M, Adlercreutz P, Mattiassion B (1992) Complex formation between chymotrypsin and ethylcellulose as a means to solubilize the enzyme in active form in toluene. *Biocatalysis* 6: 291–305

17 Virto C, Svensson I, Adlercreutz P, Mattiasson B (1995) Catalytic activity of non-covalent complexes of horse liver alcohol dehydrogenase, NAD$^+$ and polymers, dissolved or suspended in organic solvents. *Biotechnol Letts* 17: 877–882

18 Kamat SV, Beckman EJ, Russel AJ (1995) Enzyme activity in supercritical fluids. *Crit Rev Biotechnol* 15: 41–71

19 Okahata Y, Mori T (1997) Lipid-coated enzymes as efficient catalysis in organic media. *Trends Biotechnol* 15: 50–54

20 Kudryashova EV, Gladilin KA, Vakurov AV, et al. (1997) Enzyme-polyelectrolyte complexes in water-ethanol mixtures: Negatively charged groups artificially introduced into α-chymotrypsin provide additional activation and stabilization effects. *Biotechnol Bioeng* 55: 267–277

21 Kendrick BS, Meyer JD, Matsuura JE et al. (1997) Hydrophobic ion pairing as a method for enhancing structure and activity of lyophilized subtilisin BPN' suspended in isooctane. *Arch Biochem Biophys* 347: 113–118

22 Gupta MN (1995) Enzyme function in organic solvents. *In*: P Jollès, H Jörnvall (eds) *Interface between Chemistry and Biochemistry*, Birkhäuser Verlag, Basel

23 Koskinen AMP, Klibanov AM (eds). Enzymatic reactions in organic media. Blackie Academic and Professional, Glasgow 1996

24 Klibanov AM (1997) Why are enzymes less active in organic solvents than in water. *Trends Biotechnol* 15: 97–101

25 Gomez-Puyou MT, Gomez-Puyou A (1998) Enzymes in low water systems. *Crit Rev Biochem Mol Biol* 33: 53–89

26 Klibanov AM (1986) Enzymes that work in organic solvents. *CHEMTECH* 16: 354–359

27 Xu K, Griebenow K, Klibanov AM (1997) Correlation between catalytic activity and secondary structure of subtilisin dissolved in organic solvents. *Biotechnol Bioeng* 56: 485–491

28 Knubovets T, Osterhout JJ, Connolly PJ et al. (1999) Structure, thermostability, and conformational flexibility of hen egg-white lysozyme dissolved in glycerol. *Proc Natl Acad Sci USA* 96: 1262–1267

29 Schmitke, JL, Stem, LJ Klibanov, AM (1998) Organic solvent binding to crystalline subtilisin in mostly-aqueous media and in the neat solvents. *Biochem Biophys Res Comm* 248: 273–277

30 Schmitke JL, Stem LJ, Klibanov AM (1998) Comparison of X-ray crystal structures of an acryl-enzyme intermediate of subtilisin carlsberg formed in anhydrous acetonitrile and in water. *Proc Natl Acad Sci USA*. 95: 12918–12923

31 Partridge J, Dennison PR, Moore BD, Halling PJ (1998) Activity and mobility of subtilisin in low water organic media: hydration is more important than solvent dielectric. *Biochim Biophys Acta* 1386: 79–89

32 Mattiasson B, Adlercreutz P (1991) Tailoring the microenvironment of enzymes in water-poor systems. *Trends Biotechnol* 9: 394–398

33 Schulze B, Klibanov AM (1991) Inactivation and stabilization of subtilisin in neat organic solvents. *Biotechnol Bioeng* 38: 1001–1006

34 Munowitz M, Bachovchin WW, Herzfeld J et al. (1982) Acid-base and tautomeric equilibria in the solid state: ^{15}N NMR spectroscopy of histidine and imidazole. *J Am Chem Soc* 104: 1192–1196

35 Huang T-H, Bachovchin WW, Griffin RG, Dobson CM (1984) High resolution Nitrogen-15 nuclear magnetic resonance studies of α-lytic protease in solid state: Direct comparison of enzyme structure in solution and solid state. *Biochemistry* 23: 5933–5937

36 Costantino HR, Griebenow K, Langer R, Klibanov RM (1997) On the pH memory of lyophilized compounds containing protein functional groups. *Biotechnol Bioeng* 53: 345–348

37 Yang Z, Zacherl D, Russel AJ (1993) pH dependence of subtilisin dispersed in organic solvents. *J Am Chem Soc* 115: 12251–12257

38 Skrika-Alexopoulo E, Freedman RB (1993) Factors affecting enzyme characteristics of bilirubin oxidase suspensions in organic solvents. *Biotechnol Bioeng* 41: 887–893

39 Blackwood AD, Curran LJ, Moore BD, Halling PJ (1994) Organic phase buffers control biocatalyst activity independent of initial aqueous pH. *Biochim Biophys Acta* 1206: 161–165

40 Dolman M, Halling PJ, Moore BD (1997) Functionalized dendritic polybenzylethers as acid/base buffers for biocatalysis in nonpolar solvents. *Biotechnol Bioeng* 55: 278–282

41 Zacharis E, Halling PJ, Rees DG (1999) Volatile buffers can override the "pH memory" of subtilisin catalysis in organic media. *Proc Natl Acad Sci USA* 96: 1201–1205

42 Guinn RM, Skerker PS, Kavanaugh P, Clark DS (1991) Activity and flexibility of alcohol dehydrogenase in organic solvents. *Biotechnol Bioeng* 37: 303–308

43 Russel AJ, Klibanov AM (1998) Inhibitor-induced enzyme activation in organic solvents. *J Biol Chem* 263: 11624–11626

44 Stahl M, Mansson M-O, Mosbach K (1990) The synthesis of a D-amino acid ester in an organic media with α-chymotrypsin modified by a bio-imprinting procedure. *Biotechnol Lett* 12: 161–166

45 Stahl M, Ulla JW, Mansson M-O, Mosbach K (1991) Induced stereoselectivity and substrate selectivity of bio-imprinted α-chymotrypsin in anhydrous organic media. *J Am Chem Soc* 113: 9366–9368

46 Braco L, Dabulis K, Klibanov AM (1990) Production of abiotic receptors by molecular imprinting of proteins. *Proc Natl Acad Sci* USA. 87: 274–277

47 Dabulis K, Klibanov AM (1992) Molecular imprinting of proteins and other macromolecules resulting in new adsorbents. *Biotechnol Bioeng* 39: 176–185

48 Klibanov AM (1995) Enzyme memory – what is remembered and why? *Nature* 374: 596

49 Dabulis K, Klibanov AM (1993) Dramatic enhancement of enzymatic activity. *Biotechnol Bioeng* 41: 566–571

50 Khmelnitsky YL, Welch SM, Clark DS, Dordick JS (1994) Salts dramatically enhance activity of enzymes suspended in organic solvents. *J Am Chem Soc* 116: 2647–2648

51 Mingarro I, Abad C, Braco L (1995) Interfacial activation-based molecular bioimprinting of lipolytic enzymes. *Proc Natl. Acad Sci USA* 92: 3308–3312

52 Mishra P, Griebenow K, Klibanov AM (1996) Structural basis for the molecular memory of imprinted proteins in anhydrous media. *Biotechnol Bioeng* 52: 609–614

53 Griebenow K, Klibanov AM (1997) Can conformational changes be responsible for solvent and excipient effects on the catalytic behavior of subtilisin carlsberg in organic solvents. *Biotechnol Bioeng* 53: 351–362

54 Stewart NA, Taralp A, Kaplan H (1997) Imprinting of lyophilized α-chymotrypsin affects the reactivity of the active-site imidazole. *Biochem Biophys Res Commun* 240: 27–31

55 Ke T, Klibanov AM (1998) On enzymatic activity in organic solvents as a function of enzyme history. *Biotechnol Bioeng* 57: 746–750

56 Mosbach K, Ramstrom O (1996) The emerging technique of molecular imprinting and its future impact on biotechnology. *Bio/Technology* 14: 163–170

57 Soler G, Blanco RM, Fernandez-Lafuente R et al. (1995) Design of novel biocatalysts by 'Bioimprinting' during unfolding - refolding of fully dispersed covalently immobilized enzymes. *Annals NY Acad Sci* 750: 349–355

58 Russel AJ, Trudel LJ, Skipper PL et al. (1989) Antibody-antigen binding in organic solvents. *Biochem Biophys Res Commun* 158: 80–85

59 Weetall HH (1991) Antibodies in water immiscible solvents. Immobilized antibodies in hexane. *J Immunol Meth* 136: 139–142

60 Stocklein W, Gebbert A, Schmid RD (1990) Binding of triazine herbicides to antibodies in anhydrous organic solvents. *Anal Lett* 23: 1465–1476

61 Matsuura S, Hamano Y, Kita H, Takagaki Y (1993) Preparation of mouse monoclonal antibodies to okadaic acid and their binding activity in organic solvents. *J Biochem* 114: 273–278

62 Andersson LI, Nicholls IA, Mosboch K (1995) Antibody mimics obtained by non-covalent molecular imprinting. *In*: Nelson JO, Karu AE, Wong RB [eds.]. Immunoanalysis of agrochemicals: emerging technologies ACS Symposium Series 586 Washington DC: *Am Chem Soc* 89–96

63 Andersson LI (1996) Application of molecular imprinting to the development of aqueous buffer and organic solvent based radioligand binding assays for 5-propananolol. *Anal Chem* 68: 111–117

64 Janda KD, Ashley JA, Jones TM et al. (1990) Immobilized catalytic antibodies in aqueous and organic solvents. *J Am Chem Soc* 112: 8886–8888

65 Wang P, Hill TG, Warchow CA et al. (1992) New carbohydrate based materials for the stabilization of proteins. *J Am Chem Soc* 114: 378–380

66 Okahata Y, Yamaguchi M (1995) A lipid-coated catalytic antibody in water miscible organic solvents. *Tetrahedron* 51: 7673–7680

67 Nikolova P, Ward OP (1993) Whole cell biocatalysis in nonconventional media. *J Ind Microbiol* 12: 26–86

68 Nikolova P, Ward OP (1993) Effect of organic solvents on biotransformation of benzaldehyde to benzyl alcohol by free and silicon-alginate entrapped cells. *Biotechnol Tech* 7: 897–902

69 Nakamura K, Kondo S, Kawai Y, Ohno A (1993) Stereochemical control in microbial reduction XXI. Effect of organic solvents on reduction of α-ketoesters mediated by Baker's yeast. *Bull Chem Soc Jpn* 66: 2738–2743

70 Griffin DR, Yang F, Carta G, Gainer JL (1998) Asymmetric reduction of acetophenone with calcium alginate entrapped Baker's yeast in organic solvents *Biotechnol Prog* 14: 588–593

71 Andersson M, Otto R, Holmberg H, Adlercreutz P (1997) *Alcaligens entrophus* cells containing hydrogenase, a coenzyme regenerating catalyst for NADH-dependent oxidoreductases. *Biocat Biotransf* 15: 281–296

72 Andersson M, Holmberg H, Adlercreutz P (1998) Evaluation of A*lcaligens entrophus* cells as an NADH regenerating catalyst in organic-aqueous two phase systems. *Biotechnol Bioeng* 57: 79–86

73 Jaenicke R, Seckler R (1999) Spontaneous versus assisted protein folding. In: B. Bukau (ed) *Molecular chaperones and folding catalysts*. Harwood Academic Publishers, Amsterdam, 407–436

74 Rariy RV, Klibanov AM (1999) Protein refolding in predominantly organic media markedly enhanced by common salts. *Biotechnol Bioeng* 62: 704–710

75 Mason TJ (1997) Ultrasound in synthetic organic chemistry. *Chemical Society Reviews* 26: 443–451

76 Apfel RE (1997) Sonic effervescence: A tutorial on accoustic cavitation. *J Accoust Soc Am* 101: 1227–1237

77 Sinisterra JV (1992) Application of ultrasound to biotechnology: an overview. *Ultrasonics* 30: 180–185

78 Vulfson EN, Sarney DB, Law BA (1991) Enhancement of subtilisin-catalyzed interesterification in organic solvents by ultrasound irradiation. *Enzyme Microb Technol* 13: 123–126

79 Lin G, Liu H-C (1995) Ultrasound-promoted lipase-catalyzed reactions. *Tetrahedron Lett* 36: 6067–6068

80 Parker MC, Besson T, Lamare S, Legoy MD (1996) Microwave radiation can increase the rate of enzyme-catalyzed reactions in organic media. *Tetrahedron Lett* 37: 8383–86

81 Yang H, Jonsson A, Wchtje E et al. (1997) The enantiomeric purity of alcohols formed by enzymatic reduction of ketones can be improved by optimization of the temperature and by using a high cosubstrate concentration. *Biochim Biophys Acta* 1336: 51–58

82 Jonsson A, Wehtje E, Adlercreutz P (1997) Low reaction temperatures increases the selectivity in an enzymatic reaction due to substrate solvation effects. *Biotechnol Lett* 19: 85–88

**Importance of Water Activity
for Enzyme Catalysis in Non-Aqueous
Organic Systems**

Thorleif Anthonsen and Birte J. Sjursnes

Contents

1   Introduction

1.1   Significance of water for catalytic activity

The dependence of water for catalytic activity varies widely for different enzymes [1]. Some have a rather flat curve where high activity is retained over a wide range of water content, while others have bell-shaped curves and display optimum activity over only a narrow range. Furthermore, some enzymes have shown very good retention of catalytic activity under nearly anhydrous conditions. From the data it is obvious that optimising reaction conditions both with

Methods and Tools in Biosciences and Medicine
Methods in non-aqueous enzymology, ed. by M. N. Gupta
© 2000 Birkhäuser Verlag Basel/Switzerland

respect to enzyme activity and product formation requires thoughtful consideration of how water will affect enzyme activity. Further, this also reveals the necessity for a parameter connecting enzyme hydration and activity under the wide range of conditions employed.

Hydration of the enzyme provides flexibility needed for conformational changes during the catalytic process. Starting from the dry state, increasing hydration can be divided into several steps [2]. Water added interacts first with ionisable groups. This type of water is very strongly bound. Subsequently, clusters develop around charged and polar groups followed by condensation over non-polar regions giving the enzyme a monolayer of water. Regarding catalytic activity, this has been observed at hydration levels corresponding to far less than a monolayer [2–4].

When placed in organic media, there will be a competition for water among the components present. Water will be bound to the enzyme, immobilisation material, dissolve in the bulk organic phase and in the gas headspace above the reaction. The distribution of water will depend on the ability of the phases to dissolve/bind water. Especially, this will be affected by the polarity of the bulk organic phase. Adding the same amount of water to a non-polar and a polar solvent will result in different levels of enzyme hydration. Polar solvents will dissolve more water and less will be bound to the enzyme. This has often been referred to as the stripping effect [4–7]. In general, more water has to be added to polar solvents to obtain optimum catalytic activity than less polar solvents.

There are several ways to report the amount of water present in an organic system. Most commonly, total water content has been used. This is simply the sum of water added either directly or together with the different components in the mixture (like an enzyme with 10% water). Unfortunately, in relation to enzyme activity, this is not the most useful way to measure the amount of water present.

1.2 Thermodynamic water activity (a_w)

It has been shown that catalytic activity was better correlated to the amount of water actually bound to the enzyme than the total water content of the system [6]. An implication of this is that water content or concentration is not very useful for predicting catalytic activity. Thermodynamic considerations have recommended use of water activity (a_w) [8], and replotting of data confirmed correlation between enzyme activity and water activity [9]. The validity of water activity has later been demonstrated experimentally [10].

The definition of thermodynamic activity of water is conventionally given relative to pure water at the same temperature and pressure, as standard state [11]. In a gas phase it is essentially equal to the ratio of partial pressure of water vapour to that above pure water at the same temperature. Like other activities, it is by definition equal in all phases at mutual equilibrium.

Fundamentals and recommendations for using water activity have been presented [8, 12]. Experimentally, the use of a_w has been demonstrated by several studies in non-polar organic solvents for optimising of reaction conditions [10, 13–15]. Under all conditions, a similar relationship between catalytic activity and water activity was found. The polarity of the bulk organic phase will also change upon increasing substrate concentration, or when comparing different substrates. For example, increasing the concentration of butanoic acid in hexane three times would cause a significant increase in water solubility. Performing the reactions at constant water activity did however, maintain enzyme hydration at the same level [13]. Similarly, transesterification with dry pentanol, heptanol and nonanol as substrates gave very different reaction rates. When compared on basis of water activity, differences in reaction rates were greatly diminished [16]. Enzymes are often used in immobilised form on support materials. These can also affect distribution of water [17]. In most cases there was no effect on the water activity-catalytic activity profile, although some exceptions were observed [1, 18]. For reactions where water is involved, water activity will also determine the water mass action effect on the equilibrium position.

The use of water activity has no doubt proved very useful in water-immiscible solvents. In many cases, observed differences in catalytic activity have disappeared when compared on an equal water activity basis. So far, there has been little information on how water activity governs catalytic activity in water miscible solvents. One would expect that effects caused by competition for water should be equally well described. There are however questions about the direct relationship to enzyme activity. Compared to more non-polar solvents, interactions between enzyme and solvent molecules are more likely and this could affect activity in a more solvent dependent way. This aspect has been addressed and tested by replotting data for laccase catalysed reaction in various water-miscible solvents [19, 20]. Converting from water concentration to activity showed that similar optimum was found in five different alcohols. However, for dioxane, acetonitrile, tetrahydofuran and acetone, a continuing increase in optimum water activity was observed. They concluded that water activity does not always predict optimum water activity in polar solvents.

Although there still exists significant differences in absolute catalytic activity in different systems, it is strongly recommended that experiments are performed at known water activities [8]. In doing this, effects of simple competition for water between components will be accounted for, and remaining differences correctly treated. Some common methods to achieve this are described below, an overview of the matter has recently been presented [21].

1.3   Catalytic selectivity and specificity

There are several advantages of using enzymes as catalysts in chemical synthesis. Enzymes are efficient catalysts, they work under mild conditions and they are selective and specific in their catalysis. In organic chemistry the two latter properties have been particularly in focus for asymmetric synthesis and resolution of racemic mixtures respectively.

When a biocatalytic reaction does not give satisfactory results, the situation may be improved by changing the catalyst, modifying the substrate or by changing the reaction medium. The significance of these factors for resolutions by hydrolytic enzymes has been discussed [22, 23]. The medium comprises of the solvent chosen and the water content of the solvent.

While the significance of water content for catalytic activity and stability has been studied to great detail, only limited results have been published on its influence on the stereochemical properties of enzymes.

Among the eight classes of enzymes available for biocatalysis, class 3 which comprises the hydrolases, has been by far the most investigated and utilized and the discussion below refers to this class.

Asymmetric synthesis

Typical examples of enzyme catalysed enantioselective asymmetric synthesis is transesterification of a prochiral diol in organic media (Fig. 1a). High *ee*'s were obtained using vinyl acetate as acyl donor and *Pseudomonas* sp. lipase (PSL) [24]. For transesterification reactions a starting ester is needed. This is commonly called the acyl donor and it acylates the enzyme to form the acyl enzyme which in turn is attacked by a nucleophile. The nucleophile may be water (hydrolysis), an alcohol (transesterification) or an amine (amide formation). The starting material may also be a *meso* compound as in Figure 1b [25]. The influence of different hydrolases and different solvents in the transesterification of a prochiral diol has been investigated. Polar solvents like acetonitrile and nitrobenzene showed higher stereoselectivity than the less polar ones like benzene and carbon tetrachloride [26]. An explanation of the effect has been offered.

Resolution by hydrolases

Most investigations on medium effects have been carried out on resolution of racemates [27]. A racemate of a desired building block may often be produced easily by non-asymmetric synthetic methods. For instance, racemic amino acids can be obtained by the Strecker synthesis from an aldehyde, ammonia and hydrogen cyanide. When a racemate is subject to enzyme catalysis, one enantiomer reacts faster than the other and this leads to kinetic resolution. In Figure 1c is shown a racemic secondary carboxylic ester which may be resolved by a hydrolase [23]. Kinetic resolution is best characterised by the enantiomeric ratio E which is the ratio of the specificity constants of the enzyme for the two enantiomers and it is a very important parameter for a resolution process.

- Irreversible reactions, hydrolysis: In resolution there are two products called product and remaining substrate. Both products can reach very high enantiomeric excess provided E is high. The enantiomeric excesses are termed ee_p and ee_s respectively, and, as opposed to in asymmetric synth-

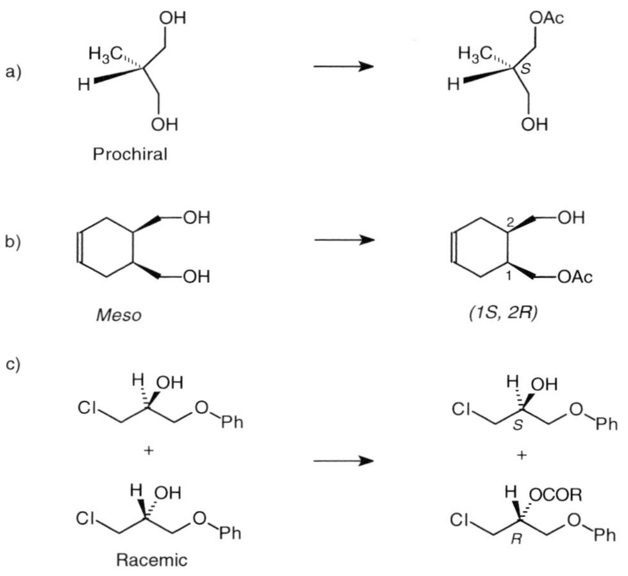

Figure 1 Production of enantiopure compounds by hydrolase catalysed reactions.

The two upper reactions are examples of asymmetric synthesis starting from a prochiral substrate (a) or a *meso*-compound (b) respectively. Example c) is kinetic resolution of a racemic mixture. In asymmetric synthesis all of the substrate may be converted into one enantiomer giving 100% yield of a product with an ee of 100%. The maximum yield of resolution is 50% of each enantiomer and their ee's depend on the degree of conversion.

esis, they depend on the degree of conversion c. Both ee_p, ee_s and c have values between 0 and 1, or 0 and 100%. For an irreversible reaction such as hydrolysis, the E-value may be calculated from either ee_p or ee_s and the degree of conversion c [28]. From Figure 2 it is evident that even for a low E-value such as 10, an ee_s close to 100% may be achieved if 25% of yield can be sacrificed. This difference between resolution and asymmetric synthesis is very important to realize. For this reason it may be easier to obtain the remaining substrate, i.e. the ester in hydrolysis, with higher ee.

- Reversible reactions, hydrolases in organic media: As mentioned above, in hydrolysis the ester may always be obtained with a high ee, but, what if the alcohol is required? This problem may be circumvented if the reaction is inverted. Instead of hydrolysis, a synthesis may be performed, an esterification or better a transesterification in non-aqueous media. Since the enzyme shows the same stereopreference no matter what is the direction of the reaction (hydrolysis or transesterification), either the alcohol or the ester may be separated as the remaining substrate. If the *(S)*-ester is the remaining substrate in hydrolysis the *(S)*-alcohol will be the remaining substrate of transesterification (Fig. 3).

A great problem with transesterification is that as the reaction proceeds, the reaction will become reversible and the equilibrium constant will become important. The enantiomer that reacts fastest in the forward direction will also react fastest in the reverse direction. The effect of this is clearly inferred from Figure 4 in which the ee_p and ee_s are calculated for a resolution with $E = 100$ and for three diferent equilibrium constants.

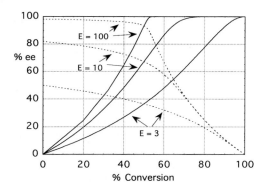

Figure 2   Enantiomeric excesses of product (ee$_p$, broken lines) and remaining substrate (ee$_s$, solid lines) plotted *versus* % conversion calculated for three different values of enantiomeric ratio E for an irreversible reaction.

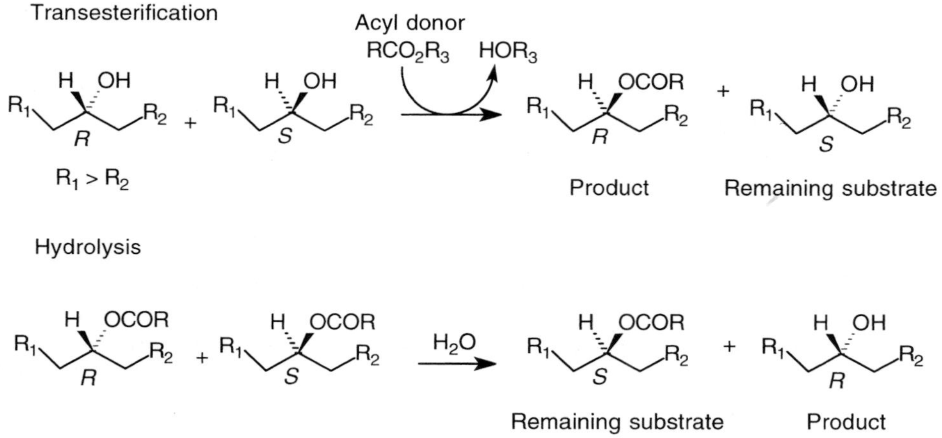

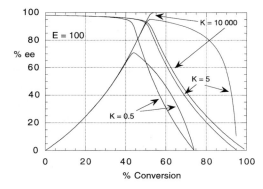

Figure 3   Comparison of the stereochemistry of products and remaining substrates in transesterification and hydrolysis.

Since the stereopreference of the enzyme is the same no matter reaction type, the *(S)*-alcohol is the remaining substrate in transesterification while the *(R)*-alcohol will be the product in hydrolysis (provided R$_1$ have higher priority than R$_2$ and both lower than OR).

Figure 4   Calculated conversion vs ee-curves for $E = 100$ and $K_{eq} = 10\,000$, 5 and 0.5 respectively.

The dramatic effect on the ee$_s$-curve is clearly visible when K_{eq} is small while the ee$_p$-curve is less severely influenced due to reversibility.

With a large K_{eq} the reaction is irreversible and the progress curve looks like the examples of Figure 2. For reactions with smaller K_{eq}-values a dramatic effect is observed for ee_s. The curve reaches a maximum, as the reaction progresses further ee_s is reduced and the curve never reaches 100% for ee_s as it always does in the irreversible case. The effect on the ee_p curve is not as dramatic, it dips down at an earlier degree of conversion. An obvious way to proceed is to push the reaction towards the product side by increasing the concentration of the reactants. Then the point where decrease of ee occurs is pushed towards higher conversion. Irreversible acyl donors such as vinyl acetate or butanoate may also be used. However, it is known that the structure of the acyl donor may influence the E-value [29].

- Significance of medium for enantiomeric excess: The medium is either a pure solvent or a mixture of solvents. Its property may significantly influence the E-value. The water content of the medium is also important. The effect of solvents and mixture of solvents for lipase catalysed resolutions has been discussed and reviewed [27]. In some cases explanations of observed effects has also been given [30,31].

The effect of water content has also been reported. In *Candida rugosa* lipase catalysed esterification of α-bromopropanoic acid a nine-fold increase in rate was obeserved when 0.125% of water was added. Under the same conditions the E-value increased by a factor of five [32]. Similarly esterfication of 2-methyl alkanoic acids using the same lipase both rate and E were enhanced when the water activity was increased to 0.76 [33, 34]. However, Pseudomonas lipases did not show a similar effect in transesterification reactions [35]. In transesterification of 1-propanol catalysed by α-chymotrypsin the E-value showed strong solvent dependence [36]. Moreover, an optimal E-value was observed in ethyl acetate with a water content of 0.3% (w/w). The competing hydrolysis reaction was found to become a problem at 0.5%.

2   Materials

Chemicals
- Butanoic acid, n-butanol, n-decane, hexanol and triethylene glycol dimethyl ether were purchased from Sigma-Aldrich.
- Hexane, diisopropyl ether and acetonitrile were of analytical grade.
- Hexyl butanoate was synthesised by refluxing hexanol (20.4 g, 0.2 moles) and butanoic acid (30.8 g, 0.35 moles) with sulphuric acid (0.6 g) in toluene (15 mL) for 10 hrs. The reaction mixture was washed with water and saturated sodium hydrogen carbonate solution, dried with anhydrous Na_2SO_4 and distilled. The resulting ester was over 98% pure (GLC).

Enzymes

- Cutinase from *Fusarium solani pisi* was kindly donated by Unilever, Vlaardingen, The Netherlands. Prior to use, cutinase was immobilised on Accurel EP100 at pH 8 with a loading of 5 mg g^{-1}.
- Lipase Type VII from *Candida rugosa* (900 units per mg) was purchased from Sigma-Aldrich.
- Immobilized lipase B from *Candida antarctica*, CALB (Novozyme 435, Novo-Nordisk A/S) used for resolutions had an activity of 7000 PLU/g, and a water content of 1–2% w/w.

Equipment

- Support: Accurel EP100 (macroporous polypropylene) was obtained from Akzo Nobel, Obernburg, Germany.
- Chiral analyses were performed using Varian 3300 and 3400 gas chromatographs equipped with CP-Chiracil-DEX-CB columns from Chrompack (25 m, 0.25 mm, 0.25 mm film density). GLC conditions, 1-phenoxy-2-propanol, temp. prog; 110 °C, hold, pressure (split) 8 (80), retention times t1 22.58, t2 23.69 resolution, Rs 1.6
- In case of CALB catalysed resolutions shaking of reaction mixtures was carried out on a G24 Environmental Incubator Shaker from New Brunswick Scientific Co, Edison, NJ USA at 30 °C and 200 strokes/min.

Solutions

- Saturated salt solutions used in this example are: LiCl, CH$_3$COOK, MgCl$_2$, K$_2$CO$_3$, Mg(NO$_3$)$_2$, KI, NaCl, KCl, KNO$_3$, K$_2$SO$_4$. These give water activities of 0.11, 0.23, 0.33, 0.43, 0.54, 0.7, 0.75, 0.85, 0.95 and 0.98 at 20°C respectively [37].
- Salt hydrates used in this example are (hydration states are shown in bracket, 0 represent the anhydrous state of the salt): Na$_2$SO$_4$ (10/0), Na$_2$HPO$_4$ (12/7), Na$_2$HPO$_4$ (7/2) and Na$_2$HPO$_4$ (2/0). These gives water activities of 0.76, 0.74, 0.57 and 0.15 at 20°C respectively. Also other salt hydrates may be used. Several have been studied systematically to examine their suitability for use in organic solvents. The available information is presented below, both for salt hydrate pairs that are recommended and those which should be avoided. For more detailed information, the work of Zacharis et al. is recommended [38].

Table 1   Recommended salt hydrate pairs for use in water control in enzyme catalysis in organic solvents. The values for a_w refer to 25°C and are taken from data compiled by Halling [53].

Salt	Hydrate pairs	a_w	References
Na_2SO_4	10/0	0.80	[13, 33, 34, 38, 52]
Na_2HPO_4	12/7	0.80	[13, 38, 62, 63]
Na_2CO_3*	10/7	0.75	[13, 50, 51, 64, 65]
Na_2HPO_4	7/2	0.62	[13, 38]
NaAc	3/0	0.28	[38, 66]
Na_2HPO_4	2/0	0.16	[13, 33]
NaI	2/0	0.12	[38, 66]

*Not to be used with acids.

Table 2   Salt hydrate pairs to be used with care.
Some of them have showed slow water transfer or other kinds of irregular behaviour in some systems, or have not been tested enough to be recommended. These should be tested before use.

Salt	Hydrate pairs	a_w	References
$Na_2B_4O_7$	10/?	0.61	[50–52]
$Na_4P_2O_7$	10/0	0.49	[13, 38, 66, 67]
$CaHPO_4$	2/0	0.50	[38]
$K_4Fe(CN)_6$	3/0	0.45	[13, 38]
$Na_2S_2O_3$	5/2	0.37	[38]
NaBr	2/0	0.35	[38]
$CuSO_4$	5/3	0.35	[38]
$Ba(OH)_2$*	8/1	0.31	[68]
Li_2SO_4	1/0	0.09	[38]

* Likely to give problems with acids.

Table 3   Salt hydrate pairs not recommended for use.

Salt	Hydrate pairs	a_w	References
Na tartrate	2/0	0.86	[69]
$ZnSO_4$	7/6	0.63	[54, 69]

3   Methods

3.1   Protocols

Protocol 1   Addition of water

1. Dry all substrates (butanoic acid and hexanol) and the organic solvent (acetonitrile) to make sure the starting point is the same for all reactions.

2. Add different amounts of water to the solvent (acetonitrile).

3. Place the solvent mixtures in vials and add substrates (butanoic acid: 100 mM, hexanol: 100 mM), immobilised lipase (100 mg to a reaction volume of 10 mL) and internal standard (triethylene glycol dimethyl ether: 25 mM).

4. Seal the vials and carry out the reactions at room temperature with shaking at 120 strokes/min.

5. Withdraw samples (100 μL). Dry samples from acetonitrile with high water content before analysis.

6. Analyse samples on GLC. Column: SGE, BP 20, 30 m, 0.32 mm i.d., 0.25 μm film thickness. Temp. program: 100°C - hold time: 1 min - 10°/min to 160°C. Increase of product is calculated from the internal standard.

7. Determine lipase activity at different water contents (initial rate or progress curve).

Protocol 2   Saturated salt solutions

1. Cover a suitable jar (not too big) with a layer of salt (type of salts, see page 21)

2. Add enough water to wet the salt. Unlike the common picture with a major water phase with only a few crystals of salt, the saturated solution should be like a wet slurry. This will ensure a saturated water phase at all times and gives an increased surface for equilibration. Some salts will form a thick syrupy slurry.

3. Place the immobilised lipase (cutinase immobilised on Accurel EP100, 100 mg) in a vial.

4. Place the vial in the jar and seal well. The arrangement is shown in Fig. 1 in ref. [39].

5. Leave the catalyst to equilibrate.

6. Repeat the procedure in a separate jar with the reaction mixture (substrates: butanoic acid and hexanol both 100 mM – organic solvent: hexane or diisopropyl ether, total volume 10 mL - internal standard: triethylene glycol dimethyl ether: 25 mM).

7. After equilibration, add the reaction mixture (10 mL to 100 mg of catalyst) to the immobilised lipase.

8. Seal the vial and carry out the reaction at room temperature with shaking at 120 strokes/min.

9. Withdraw samples (100 µL).

10. Analyse samples on GLC. Column: SGE, BP 20, 30 m, 0.32 mm i.d., 0.25 µm film thickness. Temp. program: 100°C – hold time: 1 min – 10°C/min to 160°C. The increase of product is calculated from the internal standard.

11. Determine lipase activity (initial rate or progress curve) at different water activities using different saturated salt solutions.

Protocol 3   Salt hydrates

1. Mix together organic solvent (hexane, 8 mL) and substrates (butanoic acid: 0.5 mmol, n-butanol: 1.0 mmol) and internal standard (n-decane: 0.31 mmol) in a vial.

2. Add a pair of salt hydrates (Na_2SO_4 (10/0), Na_2HPO_4 (12/7), (7/2) or (2/0), 250 mg of each).

3. Add the lipase (lipase from *Candida rugosa*, 5 mg).

4. Seal the vial and carry out the reaction at room temperature with shaking at 130 strokes/min.

5. Withdraw samples (100 µL).

6. Analyse samples on GLC. Column: J&W Scientific, DB-1701, 30 m, 0.255 mm i.d., film thickness 0.25 µm. Temp. program: 60–120°C – 10°C /min. Increase of product is calculated from internal standard.

7. Determine lipase activity (initial rate or progress curve) at different water activities using different salt hydrates.

Protocol 4   Resolution by lipase catalysed transesterification

Test reaction

1. To two 4 mL reaction vials add substrate ($2.5 \cdot 10^{-4}$ mole), solvent (3 mL) and acyl donor (vinyl acetate or vinyl butanoate or 2-chloroethyl butanoate etc., $12.5 \cdot 10^{-4}$ mole, 5 times substrate concentration).

2. The reaction is started in one vial by adding enzyme (immobilised lipase B from *Candida antarctica*, CALB, 10–20 mg). The other vial is used as control to check for non-enzyme catalysed reaction. Both vials are placed in a shaker incubator at 30 °C.

3. During the progress of the reaction 4–5 samples (50 µL each) are withdrawn from the vials using a syringe. The samples should ideally be taken over a range of 10–100% conversion. This may be difficult to achieve in the first run and a second run may be necessary. Some reactions may be very slow as they progress (after 50% conversion). To make sure that no enzyme is withdrawn from the reaction mixture, filtration through glass-wool is recommended.

4. Samples are analysed by chiral chromatography (preferably GLC for convenience) to obtain ee_s and ee_p. From these values conversion is calculated from $c = ee_s/ee_s + ee_p$. Ideally both ee_s and ee_p is obtained in the same chromatogram. If not two runs may be necessary, one for the product (ester) and one for the remaining substrate (alcohol). Sometimes the alcohol has to be esterified (acetate, trifluoro acetate).

5. Calculation of the *E*-value:
   The values of c, ee_s and ee_p are loaded into E&K calculator [40, 41].
   Specify the excess of acyl donor
   Start minimisation
   It is necessary to stop and start the minimisation several times in order to reach the true minimum.
   When the parameters minimised are not changing significantly any more, the curves corresponding to this *E* and K_{eq} value is plotted.
   The root mean square value is a good indication of how successful the minimisation was. It is also recommended to visualise the experimental point and the curve generated in the minimisation together.

Scale up of the reaction

1. The relative amount of substrates is kept the same as for the test reaction The amount of enzyme may be reduced, depending on the cost of the enzyme.

2. The reaction is followed by chiral GC until the desired enantiomeric excess of the substrate or product is reached.

3. The reaction mixture is filtered to remove enzymes.

4. The solvent is removed under reduced pressure. It is also possible to remove the acyl donor in this way, but it depends on the volatility of the substrate or product.

5. The alcohol and the ester are separated by column chromatography, or by distillation.

6. The products are characterised by optical rotation.

4   Troubleshooting

- *How can water activity be analysed?*

Coulometric Karl Fischer analysis: Even though this is not a very practical method for controlling water activity under changing conditions, it can be used for static system. Also in other connections it might be necessary to at least know the concentration of water in a relatively non-polar organic phase. Care should be taken in handling samples to avoid any change in water concentration. At low concentrations, even short exposure to surrounding atmosphere can lead to loss or gain of water, and thereby change the water activity.

Humidity sensor: An advantage is that the sensor can be placed inside the sealed reaction vial, either in the headspace or immersed in the liquid. All handling of samples is thereby avoided.

- *How can low water activity can be obtained?*

A very low water activity is difficult to obtain unless a drying agent is used. Some of them, like molecular sieves, will give low but not well-defined final water activity. If this is required, a hydrate forming salt can be used [38].

- *What if water activity in saturated salt solutions is unstable?*

To fix initial water activity, enzyme and reaction components are equilibrated in jars over saturated salt solutions as described in protocol 2. Unlike the common picture, the solution is better made like a slurry (wet salt mixture) than a liquid with only a few solid crystals. The reaction components can either be equilibrated separately or together in a reaction mixture. If equilibrated separately, the water activity may change on mixing. Most common way is to equilibrate the whole reaction mixture in one jar, and the enzyme in another. It is important to seal the jar properly to establish an internal equilibrium.

- *What if water activity after use of salt hydrates is unstable?*

Although many pairs of salt hydrates have been found to approximate ideal behaviour, there are still many that need to be tested. Note the following points:

- The pair of salt hydrates should display near ideal behaviour. Consult the literature on this point. If no data can be found, the salt has to be tested. It should give a well-defined water activity, keep this over the whole range of composition of the two hydrate forms and have short equilibration time.
- It should have no adverse effect on enzyme activity.
- It should not interfere with substrate or products. In some cases, precipitation of acids has been observed [13, 54].

5   Remarks and Conclusions

5.1   Controlled addition of water or drying

Addition of different amounts of water and measuring initial rates are often used to optimise reaction systems. The desired amount of water is commonly reported as water content. Using this relationship is simplest for water-miscible or relatively polar solvents. In these systems, most of the water used to make the mixture will be found in the liquid phase. Redistribution due to other phases like catalyst and headspace will only have minor effects on the concentration of water. Furthermore, with the substrate concentrations commonly used, these will not affect the water activity significantly. In total, the concentration of water can be based on the amount of water added, and the water activity can be found by comparing with data for the pure solvent (without substrate and catalyst).

Several methods may be used to determine the relationship between the mole fraction of water or concentration and activity. Data on aqueous-organic mixtures can be used to derive such relationships. In most cases, vapour-liquid equilibrium measurements have been used. Avoiding this part, relationships between the mole fraction of water and activity for several organic solvents have recently been published [19]. It is also possible to use theoretical models with parameters partly based on experimental measurements, like the UNIFAC group contribution method [42]. It should be noted that not all groups and interactions are covered, and deviations from experimental data have been reported [15,43, 44]. Finally, new measurements may be performed. Saturated salt solutions can be used for solvents that do not dissolve in the water phase and the water concentration in the equilibrated organic phase measured. Alternatively, water can be added and the water activity measured with a humidity sensor.

For acetonitrile there are data available on the relationship between water mole fraction and water activity [19]. This is shown in Figure 5 and 6 for mole fraction and content (%). According to this, water in the range of 44 µL to 7.8 mL was added to acetonitrile to achieve water activities from 0.12 to 0.93 (total volume: 10 mL). Taking into account the amount of water necessary to achieve different water activities, it is clear using salt hydrates is not appropriate. It would take substantial amounts of salt hydrates to reach the required water level. Pre-equilibration over saturated salt solution is not recommended either. Acetonitrile and water are completely miscible. The polar solvent will dissolve in the saturated salt solution and change the water activity.

The water activity profile is shown in Figure 7 (Protocol 1). The water activity has a marked effect on initial rate. There is no enzyme activity observed under nearly anhydrous conditions. Optimum is reach around $a_w = 0.55$. A dramatic drop follows this and no catalytic activity could be detected at water activities of 0.86 and 0.93. After examining several factors like equilibrium position, desorption of enzyme from the support and acid-base effect, the most likely cause for loss of activity at high a_w is direct effects of organic solvent. With polar solvents, direct effects of the solvent molecules on the enzyme may

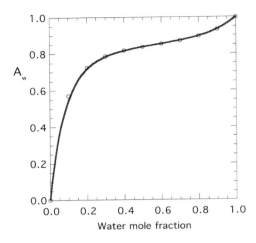

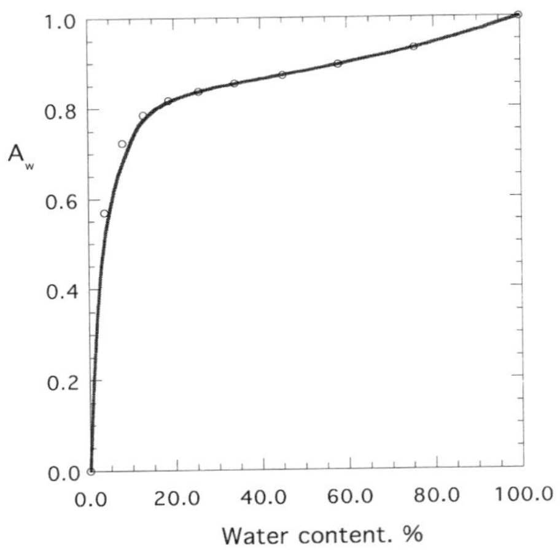

Figure 5 Shows water activity as function of mole fraction of water in acetonitrile.

Figure 6 Shows water activity as function of water content in acetonitrile.

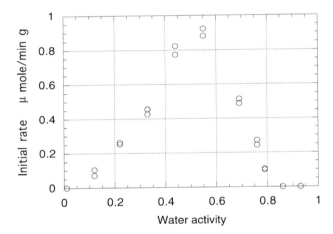

Figure 7 Initial rate vs water activity in acetonitrile for esterification of butanoic acid with hexanol catalysed by cutinase from *Fusarium solani pisi* [56–61].

Each data point is in duplicates.

change. This occurs because the concentration and activity of the organic solvent varies significantly with water content [19]. The overall trend is similar to those observed for many other enzymes under a variety of conditions.

The same procedure is not so easy for non-polar water-immiscible organic solvents. These solvents have low ability to dissolve water, the concentration of water will be low even at a water activity of 1. Other components like substrate and catalyst will greatly affect the solvation and distribution of water, and the concentration of water in the liquid phase can no longer be based on the amount of water added. Any change in substrate concentration, amount of catalyst or headspace volume will change the amount of water needed to obtain a certain water activity. This means that a relationship between water concentration and activity have to be based on exactly the same system as the reaction mixture, and will no longer be valid if for instance the substrate concentration is changed.

5.2   Saturated salt solution

The use of saturated salt solutions is an easy way to obtain an atmosphere of controlled humidity, i.e. water activity. This can be used to equilibrate reaction mixture, enzyme and in other connections like calibration of humidity sensors [45]. The saturated solution will release and absorb water at constant humidity as long as the concentration remains the same, i.e. saturated. A wide range of water activities can be obtained by this method [37].

The water activity of saturated salt solutions is fairly independent of temperature. The pressure of water will increase both over the saturated solution and over pure water with increasing temperature, and the water activity remains roughly constant. It is however important to perform the equilibrium at the same temperature as the following reaction. The amount of water needed to reach the same water activity at two temperatures will depend on the reaction mixture used.

The water activity profile for cutinase catalysed esterification of butanoic acid with hexanol in hexane and diisopropyl ether is shown in Figure 8 (Protocol 2). The absolute rates are different in the two solvents, but the profiles are similar. This illustrates the benefit of doing this type of comparison on a water activity basis. The water content at equal water activities will be very different.

Saturated salt solutions may also be used for controlling water activity during reaction. One possibility is to leave the reaction vial open in a closed jar with saturated salt solution [46, 47]. As the reaction proceeds, the water activity will be controlled by equilibration through the atmosphere. To obtain a constant water activity, the rate of equilibration between the reaction mixture and atmosphere has to be higher than the change of water activity in the mixture. By pumping a saturated salt solution through silicon tubing immersed in the reaction mixture the water activity may be kept at a constant level [48]. The water can permeate the wall of the tubing, and the system is equilibrated to the same water activity as the saturated salt solution. This system also offers

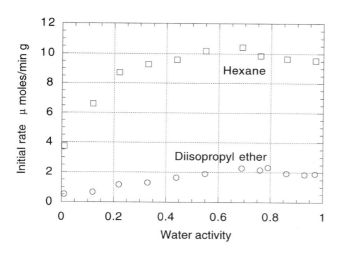

Figure 8   Initial rate vs water activity in hexane and diisopropyl ether for esterification of butanoic acid with hexanol catalysed by cutinase from *Fusarium solani pisi.*

an easy way of changing the water activity during reaction by simply changing the saturated salt solution pumped through [49]. This method may for example be used to shift an equilibrium position at the end of a reaction at the expense of lower catalytic activity. A water activity corresponding to optimum catalytic activity is used until equlibrium is reached. Then the water activity is decreased (i.e. for esterification) or increased (i.e. for hydrolysis) to obtain maximum yield of product.

5.3 Salt hydrates

The use of salt hydrates offers a very easy and convenient method for controlling water activity. It requires no special equipment, and a range of water activities can be tested very quickly. One of the major drawbacks of the method is the presence of two solid components, salt and enzyme. The reuse of the enzyme would require separation of the salt hydrates.

A simple and convenient method for controlling water activity during reaction is to add a pair of solid salt hydrates (same type of salt, but different hydration level) [13, 50–52]. Ideal behaviour implies a fixed equilibrium water pressure and hence a_w. The lower hydrate or anhydrous form is able to absorb water while the higher can release water. The pair will act as a water buffer keeping the water activity constant. The buffering ability of solid salt hydrates is illustrated in Figure 9. The three hydrate forms of disodium hydrogenphosphate are able to keep the water activity of a system at three fixed values. The ability of salt hydrates to maintain and control water activity has been demonstrated by changing reaction conditions like type of solvent, substrate concentration and amount of enzyme [13]. The theory and data on possible salt hydrates have been published [53] followed by recommendations on suitable salt hydrates [38].

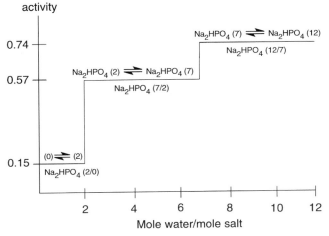

Figure 9 Buffering ability of solid salt hydrates.

The three hydrate forms of disodium hydrogenphosphate are able to keep the water activity of a system at three fixed values.

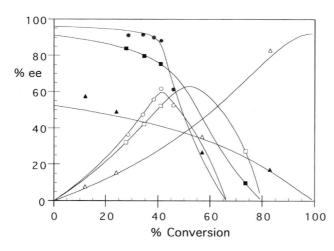

Figure 10   Resolution of 1-phenoxy-2-propanol with lipase B from Candida antarctica and vinyl butanoate in hexane with a_w at 0, (triangle) 0.65 (square) and 0.85 (circle). Open symbols: ee-value of substrate. Closed symbols: ee-value of product. The E-values were calculated to be 4, 21 and 48 respectively, i.e. increasing with increasing water activity. The equilibrium constants K_{eq} were calculated to be ∞, 0.75 and 0.30 respectively.

In protocol 3, four pairs of salt hydrates were used to find the best conditions for esterification of butanoic acid with butanol catalysed with lipase from *Candida rugosa*. The effect on initial rate is shown in Figure 6 in ref. [13]. The conversion was in all cases close to 100%. Highest initial rate was obtained for reactions with Na_2SO_4 (10/0), $a_w = 0.76$ and Na_2HPO_4 (12/7), $a_w = 0.74$ added. As the water activity decreases, the initial rate decreases.

One of the major advantages with salt hydrates is that they maintain the desired water activity when the composition of the reaction mixture is changed. The data when substrate concentration was increased 3 times or when enzyme amount was increased from 1.5 to 30 mg is available. Reaction rates without salt hydrates were shown for comparison [13].

In theory, an ideal pair of salt hydrates will control water activity as long as both forms are present. The relative amounts make no difference. It may even not be necessary to add both hydrate forms of a salt. The lower hydrate may be produced in sufficient amounts by initial water migration to other (dry) components of the reaction mixture, while a higher can be formed if other components are initially relatively wet. If the amount of salt added is not important and does not create practical problems, the easiest way to proceed is simply to add excess of both hydrate forms. This will in principle have no effect either way. In case the minimum amount of salt hydrates is required, this can be estimated from a water budget for the system [13]. Sources and sinks of water to be considered is water dissolved in the organic phase, associated with the catalyst, in the head space, formed or consumed during reaction and gained/lost through exchange with the surrounding atmosphere during handling. For kinetic experiments it is advisable to pre-equilibrate to make sure correct a_w is achieved. Significant differences in initial rates have been observed for non- and pre-equilibrated reactions [38].

With increasing temperature, the equilibrium water vapour pressure of the salt hydrate pair will increase. This is also the case for pure water, however, the increase is in most cases greater for the salt hydrates giving a net increase in water activity with increasing temperature. It is especially important to be aware of the maximum temperature limit for many salts. Above this temperature, the salt hydrates will turn into solution, and the water activity will change.

5.4   Significance of water activity for E and K_{eq}

In transesterification of 1-phenoxy-2-propanol using lipase B from *Candida antarctica* (CALB) and 2-chloroethyl butanoate as acyl donor the *E*-value increased significantly when a_w was increased to 0.61 [40] (Protocol 4). However, the ee_s and ee_p curves showed the behaviour illustrated in Figure 4. Even when the irreversible acyl donor vinyl butanoate was used, the damaging effect of reversibility was observed (Figure 10) [23]. For water activity values of 0, 0.65 and 0.85 the corresponding *E*-values were 4, 21 and 48 respectively. The equilibrium constants K_{eq} were calculated to be 8, 0.75 and 0.30 respectively. The significance of these values are, however, not clear.

During a transesterification in nonaqueous organic media two equilibria, with identical K_{eq} have to be considered. When water is present in the reaction system, three additional equilibria are involved (Figure 11). Hence the calculated equilibrium constant involves all five reactions. What is important from a practical viewpoint is the ee vs degree of conversion curves which show that ee_s is reduced beyond a certain *c*-value.

Recent results for transesterification of 1-bromo-3-phenoxy-2-propanol catalysed by lipase B from *Candida antarctica* using vinyl butanoate as acyl donor in different solvents and at different a_w indicate that solvents which show different *E*-values when a_w is zero, become more similar when a_w is increased (Table 4) [55]. Solvents such as dioxane, hexane, benzene and toluene which give very different *E*-values in dry state become almost equal at a_w = 0.65. It is interesting to notice that while *E* is decreased for dioxane it is increased for toluene. This accord with previous reports that have shown that increased a_w may influence *E* in both directions (See part 1.3.2). This illustrates that not only the solvent, but also the water activity is essential for the medium properties in enzyme catalysed reactions.

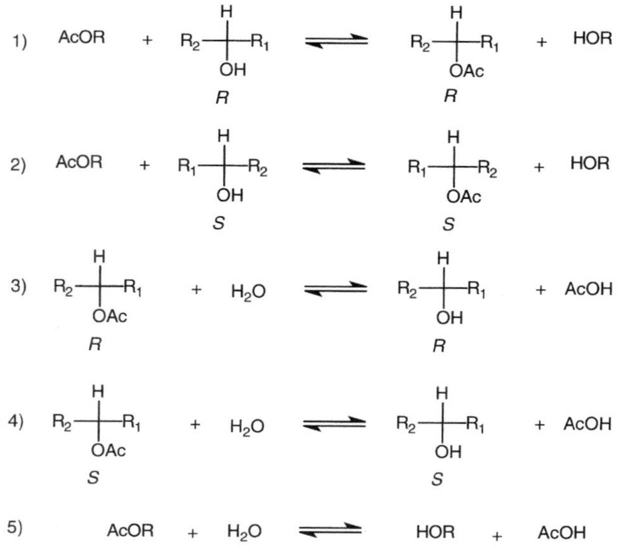

Figure 11 Five equilibria which comprise the reaction system in transesterification when water is present.

Equilibria 1 and 2, which are the reactions between the acyl donor (AcOR) and the two enantiomeric alcohols obviously have the same equilibrium constant. This is also true for the equilibria between the formed enantiomeric acetates and water. The 5th equilibrium which has to be considered is the hydrolysis of the acyl donor which do not have an enantiomeric participant.

Table 4 *E*-values for transesterification of 1-bromo-3-phenoxy-2-propanol catalysed by lipase B from *Candida antarctica* using vinyl butanoate as acyl donor in different solvents and at different water activities.

Solvent	$a_w \approx 0$	$a_w = 0.177$	$a_w = 0.65$
Dioxane	93	69	57
Hexane	58	42	55
Benzene	23	35	59
Toluene	14	22	57

References

1 Valivety R, Halling P, Peilow A, Macrae A (1992) Lipases from different sources vary widely in dependence of catalytic activity on water activity. *Biochim. Biophys Acta* 1122: 143–146

2 Rupley J, Graton E, Careri G (1983) Water and globular proteins. *Trends Biochem. Sci.* 8: 18–22

3 Rupley J, Yang P-H, Tollin G (1980) Thermodynamic and related studies of water interacting with proteins. *ACS Symp. series* 127: 111–132

4 Zaks A, Klibanov A (1988) Enzymatic catalysis in nonaqueous solvents. *J. Biol. Chem.* 263: 3194–3201

5 Zaks A, Klibanov A (1985) Enzyme-catalyzed processes in organic solvents. *Proc. Natl. Acad. Sci. USA* 82: 3192–3196

6 Zaks A, Klibanov A (1988) The effect of water on enzyme action in organic media. *J. Biol. Chem.* 263: 8017–8021

7 Gorman L, Dordick J (1992) Organic solvents strip water off Enzymes. *Biotechnol. Bioeng.* 39: 392–397

8 Halling P (1994) Thermodynamic predictions for biocatalysts in nonconventional media: Theory, test and recomendations for experimental design and analysis. *Enzyme Microb. Technol.* 16: 178–206

9 Halling P (1990) High affinity binding of water by proteins is similar in air and in organic solvents. *Biochim. Biophys. Acta* 1040: 225–228

10 Valivety R, Halling P, Macrae A (1992) Reaction rate with lipase catalyst shows similar dependence on water activity in different organic solvents. *Biochim. Biophys. Acta* 1118: 218–222

11 Eisenberg D, Crothers D (1979) *Physical Chemistry with Applications to the Life Sciences*. The Benjamin/Cummins Publishing Co 297–300

12 Bell G, Halling P, Moore B, Partridge J, Rees D (1995) Biocatalyst behaviour in low-water systems. *TIBTECH* 13: 468–473

13 Kvittingen L, Sjursnes B, Anthonsen T, Halling P (1992) Use of salt hydrates to buffer optimal water level during lipase catalysed synthesis in organic media: A practical procedure for organic chemists. *Tetrahedron* 48: 2793–2802

14 Martins J, DeSampaio T, Carvalho I, Barreiros S (1994) Lipase catalyzed esterification of glycidol in nonaqueous solvents: Solvent effects on enzymatic activity. *Biotechnol. Bioeng.* 44: 119–124

15 DeSampaio T, Melo R, Moura T, Michel S, Barreiros S (1996) Solvent effects on the catalytic activity of subtilisin suspended in organic solvents. *Biotechnol. Bioeng.* 50: 257–264

16 Goldberg M, Thomas D, Legoy M-D (1990) The control of lipase-catalysed transesterification and esterification reaction rates. Effects of substrate polarity, water activity and water molecules on enzyme activity. *Eur. J. Biochem.* 190: 603–609

17 Reslow M, Adlercreutz P, Mattiasson B (1988) On the importance of the support material for bioorganic synthesis. Influence of water partition between solvent, enzyme and solid support in water-poor reaction media. *Eur. J. Biochem.* 172: 573–578

18 Adlercreutz P (1991) On the importance of the support material for enzymatic synthesis in organic media. Support effects at controlled water activity. *Eur. J. Biochem.* 199: 609–614

19 Bell G, Janssen A, Halling P (1997) Water activity fails to predict critical hydration level for enzyme activity in polar organic solvents: Interconversion of water concentrations and activities. *Enzyme Microb. Technol.* 20: 471–477

20 van Erp S, Kamenskaya E, Khmelnitsky Y (1991) The effect of water content and nature of organic solvent on enzyme activity in low water media. A quantitative description. *Eur. J. Biochem.* 202: 379–384

21 Bell G, Dolman M, Halling P, Hessbrügge B, Moore B, Partridge J, Rees D, Vadya A (1998) Practical methods for measurement and control of water in non-aqueous biocatalysis. *Prep. Biotrans.* Submitted

22 Faber K, Ottolina, G., and Riva, S. (1993) Selectivity-enhancement of hydrolase reactions. *Biocatalysis* 8: 91–132.

23 Anthonsen T, Hoff B (1998) Resolution of derivatives of 1,2-propanediol with lipase B from *Candida antarctica*. Effect of substrate structure, medium, water activity and acyl donor on enantiomeric ratio. *Chem. Phys. Lipids* 93: 199–207

24 Santianello E, Ferraboschi P, Grosenti P (1990) An efficient chemo-enzymatic approach to the enantioselective synthesis of 2-methyl-1,3-propanediol derivatives. *Tetrahedron Lett.* 31: 5657–5660

25 Ader U, Breitgoff D, Klein P, Laumen KE, Schneider MP (1989) Enzymatic ester hydrolysis and synthesis-Two approaches to cycloalkane derivates of high enantiomeric purity. *Tetrahedron Lett.* 30: 1793–1796

26 Terradas F, Teston-Henry M, Fitzpatrick P, Klibanov A (1993) Marked Dependence of Enzyme Prochiral Selectivity on the Solvent. *J. Am. Chem. Soc.* 115: 390–396

27 Anthonsen T, Jongejan J (1997) Solvent effect in lipase catalysed racemate resolution. *Meth. Enzymol.* 286: 473–495

28 Chen C-S, Fujimoto Y, Girdaukas G, Sih CJ (1982) Quantitative analyses of biochemical kinetic resolutions of enantiomers. *J. Am. Chem. Soc.* 104: 7294–7299

29 Hoff BH, Anthonsen HW, Anthonsen T (1996) The enantiomer ratio strongly depends on the alkyl part of the acyl donor in transesterification with lipase B from *Candida antarctica*. *Tetrahedron: Asymmetry* 7: 3187–3192

30 Fitzpatrick P, Klibanov A (1991) How can the solvent affect enzyme enantioselectivity? *J. Am. Chem. Soc.* 113: 3166–3171

31 Lundhaug K, Overbeeke PLA, Jongejan J, Anthonsen T (1998) Organic co-solvents restore the inherently high enantiomeric ratio of lipase B from *Candida antarctica* in hydrolytic resolution by relieving the enatiospecific inhibition of product alcohol. *Tetrahedron: Asymmetry* 9: 2851–2856

32 Kitaguchi H, Itoh I, Ono M (1990) Effects of water and water-mimicking solvents on the lipase-catalyzed esterification in apolar solvents. *Chem. Lett.* 7: 1203–1206

33 Högberg H, Edlund H, Berglund P, Hedenström E (1993) Water activity influences enantioselectivity in a lipase-catalysed resolution by esterification in an organic solvent. *Tetrahedron: Asymmetry* 4: 2123–2126

34 Berglund P, Vörde C, Högberg H-E (1994) Esterification of 2-methylalkanoic acids catalyzed by lipase from *Candida rugosa* - Enantioselectivity as a function of water activity and alcohol chain-length. *Biocatalysis* 9: 123–130

35 Bovara R, Carrea G, Ottolina G, Riva S (1993) Water Activity does not Influence the Enantioselectivity of Lipase PS and Lipoprotein Lipase in Organic Solvents. *Biotechnol. Letters* 15: 169–174

36 Kawashiro K, Sugahara H, Sugiyama S, Hayashi H (1995) Effects of water content on the enantioselectivity of α-chymotrypsin catalysis in organic media. *Biotechnology Lett.* 17: 1161–1166

37 Greenspan L (1977) Humidity fixed points of binary saturated aqueous solutions. *J. Res. Nat. Bur. St. Phys. Chem.* 81A: 89–96

38 Zaccharis E, Omar I, Partridge J, Robb D, Halling P (1997) Selection of salt hydrate pairs for use in water control in enzyme catalysis in organic solvents. *Biotechnol. Bioeng.* 55: 367–374

39 Hutcheon G, Halling P, Moore B (1997) Measurement and control of hydration in nonaqueous biocatalysis. *Meth. Enzymol.* 286: 465–472

40 Anthonsen HW, Hoff BH, Anthonsen T (1996) Calculation of enantiomer ratio and equilibrium constants in biocatalytic ping-pong bi-bi resolutions. *Tetrahedron: Asymmetry* 7: 2633–2638

41 Anthonsen H (1996–7) E&K calculator version 2.03. *http://Bendik@kje.ntnu.no*

42 Fredenslund A, Gmehling J, Rasmussen P (1977) *Vapor-liquid equilibria using Unifac*. Elsevier, Amsterdam

43 van Tol J (1994) Description of the kinetics and enantioselectivity of lipases in various media. *Doctoral Thesis, Delft University of Technology, The Netherlands*

44 Janssen AE, Padt Avd, Sonsbeek H, Riet Kvt (1993) The effect of organic solvents on the equilibrium position of enzymatic acylglycerol synthesis. *Biotechnol. Bioeng.* 41: 95–103

45 Goderis H, Ampe G, Feyten M, Fouwe B, Guffens W, Cauwenbergh S v, Tobback P (1987) Lipase-catalyzed ester exchange reactions in organic media with controlled humidity. *Biotechnol. Bioeng.* 30: 258–266

46 Bloomer S, Adlercreutz P, Mattiasson B (1991) Triglyceride interesterification by lipases 2. Reaction parameters for the

reduction of trisaturated impurities and diglycerides in batch reactions. *Biocatalysis* 5: 145–162

47 Ljunger G, Adlercreutz P, Mattiasson B (1994) Enzymatic synthesis of octyl-ß-glucoside in octanol at controlled water activity. *Enzyme Microb. Technol.* 16: 751–755

48 Wehtje E, Svensson I, Adlercreutz P, Mattiasson B (1994) Continuous control of water activity during biocatalysis in organic media. *Biotechnol. Techniques* 7: 873–878

49 Svensson I, Wehtje E, Adlercreutz P, Mattiasson B (1994) Effects of water activity on reaction rates and equilibrium positions in enzymatic esterifications. *Biotechnol. Bioeng.* 44: 549–556

50 Kuhl P, Halling P (1991) Salt hydrates buffer water activity during chymotrypsin-catalyzed peptide synthesis. *Biochim. Biophys. Acta* 1078: 326–328

51 Kuhl P, Eichhorn U, Jakubke H (1991) Alpha-Chymotrypsin catalyzed synthesis of peptides in hexane using salt hydrates. *Pharmazie* 46: 53–58

52 Kuhl P, Eichhorn U, Jakubke H (1992) Thermolysin and chymotrypsin-catalysed peptide synthesis in the presence of salt hydrates. In: J Tramper, M Vermue, H Beeftink, U v Stockar (eds): *Biocatalysis in non-conventional media.* Elsevier, Amsterdam, 513–518

53 Halling P (1992) Salt hydrates for water activity control with biocatalysts in organic media. *Biotechnol. Techniques* 6: 271–276

54 Sjursnes B, Kvittingen L, Anthonsen T, Halling P (1992) Control of water activity by using salt hydrates in enzyme catalysed esterification in organic media. In: J Tramper, M Vermue, H Beeftink, Uv Stockar (eds): *Biocatalysis in non-conventional media.* Elsevier, Amsterdam, 451–457

55 Rønstad A, Hoff B, Anthonsen T (1998) *Unpublished results*

56 Martinez C, DeGeus P, Lauwereys M, Matthyssen G, Cambillau C (1992) *Fusarium solani* cutinase is a lipolytic enzyme with a catalytic serine accessible to solvent. *Nature* 356: 615–618

57 Sebastio M, Cabral J, AiresBarros M (1993) Synthesis of fatty-acid esters by a recombinant cutinase in reversed micelles. *Biotechnol. Bioeng.* 42: 326–332

58 Osman S, Gerard H, Moreau R, Fett W (1993) Model substrates for cutinases. *Chem. Phys. Lipids* 66: 215–218

59 Martinez C, Nicolas A, van Tilbeurgh H, Egloff M-P, Cudrey C, Verger R, Cambillau C (1994) Cutinase, a lipolytic enzyme with a preformed oxyanion hole. *Biochemistry* 33: 83–89

60 Mannesse M, Cox R, Koops B, Verheij H, de Haas G, Egmond M, van der Hijden H, deVlieg J (1995) Cutinase from *Fusarium solani pisi* hydrolyzing triglyceride analogues. Effect of acyl chain length and position in the substrate molecule on activity and enantioselectivity. *Biochemistry* 34: 6400–6407

61 Flipsen J, van der Hijden H, Egmond M, Verheij H (1996) Action of cutinase at the triolein-water interface. Characterisation of interfacial effects during lipid hydrolysis using the oil-drop tensiometer as a tool to study lipase kinetics. *Chem. Phys. Lipids* 84: 105–115

62 Yang Z, Robb D, Halling P (1992) Variation of tyrosinase activity with solvent at constant water activity. In: J Tramper, M Vermue, H Beeftink, U v Stockar (eds): *Biocatalysis in non-conventional media.* Elsevier, Amsterdam,

63 Rosell C, Vaidya A (1995) Twin core packed-bed reactors for organic-phase enzymatic esterification with water activity control. *Applied Microbiol. Biotechnol.* 44: 283–286

64 Kuhl P, Posselt S, Jakubke H (1981) Alpha-chymotrypsin-katalysierte synthese von tripeptidamiden im wasser-organischen zweiphasensystem. *Pharmazie* 36: 463–465

65 Kuhl P, Halling P, Jakubke H-D (1990) Chymotrypsin suspended in organic solvents with salt hydrates is a good catalyst for peptide synthesis from mainly undissolved reactants. *Tetrahedron Lett.* 31: 5213–5216

66 Han J, Rhee J (1998) Effect of salt hydrate pairs for water activity control on lipase-catalyzed synthesis of lysophospholipids in a solvent-free system. *Enzyme Microb. Technol.* 22: 158–164

67 Yang Z, Zacherl D, Russel A (1993) pH-Dependence of subtilisin dispersed in organic-solvents. *J. Am. Chem. Soc.* 115: 12251–12257

68 Kim J, Han J, Yoon J, Rhee J (1998) Effect of salt hydrate pair on lipase-catalyzed regioselective monoacylation of sucrose. *Biotechnol. Bioeng.* 57: 121–125

69 Yang Z, Robb D (1993) Use of salt hydrates for controlling water activity of tyrosinase in organic solvents. *Biotechnol. Techniques* 7: 37–42

Engineering of Enzymes via Immobilization and Post-Immobilization Techniques: Preparation of Enzyme Derivatives with Improved Stability in Organic Media

Gloria Fernández-Lorente, Roberto Fernández-Lafuente, Pilar Armisén, Pilar Sabuquillo, Cesar Mateo and José M. Guisán

Contents

Methods and Tools in Biosciences and Medicine
Methods in non-aqueous enzymology, ed. by M. N. Gupta
© 2000 Birkhäuser Verlag Basel/Switzerland

1   Introduction

The high potential of enzyme biotransformations in nonaqueous media has been adequately emphasised throughout the present volume. However, enzymes also have important limitations for working in non-aqueous media. For example, enzymes are usually inactivated in the presence of high concentrations of organic cosolvents or in the presence of hydrophobic interfaces of non-miscible solvents and so on [1]. Such limitations may not be very relevant when working at laboratory scale (e.g. by performing a short unique biotransformation) but they can become critical when trying to scale up such exciting biotransformations to an industrial scale (e.g. trying to perform a number of long reaction cycles).

Enzyme biotransformations in organic media are usually performed by immobilized derivatives. At first glance, immobilization (e.g. covalent attachment on solid supports) is mainly used to improve the re-use or continuous use of industrial enzyme. However, immobilization can be associated with some interesting stabilizing effects (see Fig. 1). For example, because of full dispersion, immobilized enzyme molecules are not able to aggregate in the presence of organic cosolvents [2]. On the other hand, immobilization inside porous structures also prevents enzyme molecules from interaction with hydrophobic interfaces present in the reaction media. When working at high water activities, solvents are not able to penetrate inside the porous structure of the support (because of capillary phenomena) and hence the porous structure where the enzyme is immobilized retains the aqueous environment [2, 3]. So, enzyme-solvent interactions will only occur when working at very low water activity. Under these conditions, solvents are forced to penetrate inside the porous structure of the support and hence come in contact with the enzyme molecules.

In spite of such interesting stabilizing effects, random covalent immobilization of enzymes (e.g. one-point covalent attachment on solid supports) should

- **aggregation is now impossible**

- **deleterious interaction with solvent interfaces
  is now impossible**

Figure 1   Stabilizing effects of random covalent immobilization of enzymes.

not promote any significant increase in the conformational rigidity of immobilized enzyme molecules [2]. In these cases, immobilized enzyme molecules should be as sensitive as soluble enzymes against conformational changes promoted by organic solvents.

In this communication we propose a dual-strategy that strongly improves the conformational stability of immobilized enzymes against the presence of organic solvents.

1.1 Directed immobilization: rigidification of enzymes via multipoint covalent immobilization on pre-existing supports

The involvement of a number of enzyme residues in its covalent immobilization on the solid support could generate important stabilizing effects [4, 5]. If these residues were attached to a rigid support through very short spacer arms, the relative distances between such a number of enzyme residues would have to be preserved during any conformational change undergone by the enzyme molecule. In this way, conformational changes promoted by organic solvents on such multipoint immobilized enzyme molecules should be now greatly reduced [6] (Fig. 2).

Obviously, the practical performance of such a simple hypothesis is not very easy. Enzymes and solid supports are neither similar nor complementary structures and it is difficult to design such intense enzyme-support multipoint immobilizations. Such a goal has been a key point in the work of our laboratory. The immobilization of enzymes on supports having large internal surfaces and activated with very dense layers of glyoxyl groups (Support - CH_2-CHO) has been proposed as a very interesting immobilization-stabilization system. The main and unique features of such a system are shown in Figures 3 and 4 and they have been previously discussed [4, 5, 7, 8].

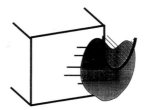

Distances among residues involved in immobilization have to be preserved during any conformational change induced by any denaturing agent

Figure 2   Rigidification of enzymes by multipoint covalent immobilization.

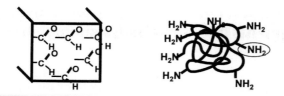

● "geometrical congruence" enzyme / support

● very highly activated supports

● very long multi-interactions

● immobilization through the area containing the highest density of amino groups

● absence of steric hindrances for intramolecular multi-interactions

● minimal chemical modifications

Figure 3 Multipoint immobilization of enzymes on glyoxyl-agarose.

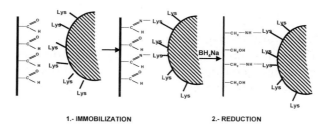

1.- IMMOBILIZATION   2.- REDUCTION

- only the enzyme residues involved in multipoint immobilization are modified
- chemical modification of such residues is very mild: from primary into secondary amino groups

Figure 4 The chemistry of multipoint immobilization of enzymes on glyoxyl-supports

1.2   Post-immobilization techniques: creation of hydrophilic nano-environments fully surrounding immobilized enzyme molecules

The generation of hydrophilic environments very near to the enzyme surface may promote some interesting stabilizing effects against the presence of organic solvents [9]:

i) the enzyme surface is prevented from direct interaction with a harmful medium (high concentration of cosolvents, non-miscible solvent interfaces, etc.); ii) some hydrophobic pockets on the enzyme surface might be now masked by the hydrophilic environment; iii) the real concentration of organic cosolvent in the enzyme surrounding might be reduced and even a slight reduction of such a concentration (e.g. from 50% down to 25%) might greatly improve the stability of the enzyme against these very distorting agents [8].

We have proposed a 3-step strategy to get a very intense hydrophilization of immobilized enzyme molecules (Fig. 5):

- immobilization on highly hydrophilic supports having large internal surfaces. In this way, the selection of highly hydrophilic supports to perform the multipoint covalent immobilization should provide a partial hydrophilic nano-environment surrounding the area of the enzyme that is directly in contact with the support.
- further covalent immobilization of highly hydrophilic polymers (e.g. poly-ethyleneimine) on the same large surfaces of the solid support surface should provide an additional hydrophilic environment surrounding all lateral areas of each immobilized enzyme molecule.
- the chemical modification of the enzyme-polyamine derivative with hydrophilic polyfunctional macromolecules (e.g. aldehyde-dextran that can be further converted into a poly-hydroxyl structure) [9] should complete the hydrophilic surrounding of all the surface of each immobilized enzyme molecule [9].

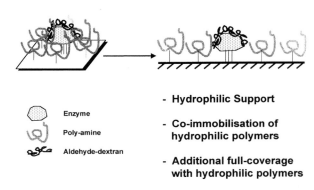

Enzyme

Poly-amine

Aldehyde-dextran

- **Hydrophilic Support**

- **Co-immobilisation of hydrophilic polymers**

- **Additional full-coverage with hydrophilic polymers**

Figure 5   Hydrophilic nano-environments fully surrounding immobilized enzymes.

1.3   Immobilization-stabilization of Pig Liver Esterase (PLE)

The applicability of such a strategy for stabilization of enzymes will be exemplified by using PLE as an interesting industrial enzyme. Pig liver esterase (E.C. 3.1.1.1) is a serine type of hydrolase that catalyzes the stereoselective hydrolysis of a wide variety of esters. Many PLE substrates are highly hydrophobic and the use of high concentrations of them requires the use of water-cosolvent mixtures. On the other hand, it has been observed that stereoselectivity can be enhanced by using such water-cosolvent mixtures [10]. PLE was also used to catalyze transesterifications of chiral alcohols in biphasic organic-aqueous systems having high volume fractions of the organic phase [10]. In this way, the preparation of PLE derivatives with very enhanced stability towards organic media may be relevant to their use in such interesting synthetic biotransformations.

2   Materials

Chemicals

Amino agarose	Hispanagar
Crosslinked 6% agarose	Hispanagar
Glutaraldehyde (25%, w/v)	Merck
Glycidol	Sigma
Glyoxyl agarose	Hispanagar
P-nitrophenylpropionate (pNPP)	Sigma
Polyethleneimine (MW 25,000)	Aldrich

Any other reagents used were of analytical grade.

Enzymes

Esterase from pig liver (PLE)	Sigma

3   Methods

3.1 Protocols

Protocol 1   Activation of Agarose gels

A   Preparation of poorly activated glutaraldehyde-agarose

Cross-linked 6% agarose gels were utilized as support for enzyme immobilization. Such supports are very resistant to strong mechanical stirring as well as to any other gentle stirring device. However, agarose gels are very sensitive to strong magnetic stirring.

These crosslinked gels already have up to 20 µmoles of glycidyl residues per millilitre of packed agarose. These easily activable groups are formed as a result of a side-reaction during the crosslinking of agarose gels with epichlorohydrin. In this way, very poorly activated gels can be easily prepared. To prepare glutaraldehyde-agarose activated with 15 µmoles per mL of agarose, the following protocol was used.

A1   Preparation of glyoxyl agarose

100 mL of agarose was suspended in 900 mL of distilled water containing 1,500 µmoles of sodium periodate. The suspension was maintained under very gently stirring for 2 hrs, then the gel was filtered (e.g. using a Buchner flask with glass-sintered funnel connected to a vacuum line) and washed with excess of distilled water.

A2   Preparation of amino-agarose

100 mL of glyoxyl-agarose were suspended in 900 mL of 2.0 M ethylenediamine at pH 10 and the suspension was gently stirred for 2 hrs at room temperature. Then, 10 g of solid sodium borohydride was added in order to reduce the amine-aldehyde bonds (Schiff's bases) as well as to reduce any nitro group produced by undesirable side oxidation of amino groups. Reduction was performed for 2 hrs. The amino-agarose gel was filtered and washed 4 times with 3 volumes of basic buffer (0.1 M sodium borate, 1 M NaCl, pH 9.0), with 3 volumes of acidic buffer (0.1 M sodium acetate, 1 M NaCl, pH 4.0) and with a large excess of distilled water. Through this treatment all glyoxyl groups were transformed into primary amino groups having a very low pK value of around 7 [11].

A3   Preparation of glutaraldehyde-agarose

100 mL of amino-agarose were suspended in 900 mL of 1.1 M glutaraldehyde in 0.1 M sodium phosphate at pH 7. The suspension was very gently stirred for 14 hrs and then the activated support was filtered (e.g. as described above) and washed with a great excess of water. This treatment assured the total conversion of the amino groups to glutaraldehyde groups. Therefore, the final concentration of glutaraldehyde residues is determined by the initial concentration of glyoxyl groups in the support that can be easily controlled by controlling the amount of periodate used in the first oxidation of cross-linked agarose.

B   Preparation of highly activated glyoxyl-agarose

In this case, the total conversion of all primary hydroxyl groups placed on the internal surface of agarose gels into glyceryl groups is intended. For this purpose, the following protocol has been designed [4]:

B1   Activation of agarose gel to glyceryl-agarose

100 mL of agarose was suspended in distilled water up to a total volume of 120 mL and the suspension is cooled in a ice bath. Then, 34 mL of 1.7 N NaOH solution containing 0.95 g of borohydride was added and the new suspension was very gently stirred. Thereafter, 24 mL of glycidol were added to the suspension. Addition was made very slowly (aliquots of 2 mL per minute) in order to prevent an undesirable increase in temperature. After 2 hrs in the ice bath, the suspension was left to reach the room temperature and the activation is continued for 16 hrs. Then, the activated support was washed with a large excess of distilled water.

B2   Activation of glyceryl-agarose to glyoxyl-agarose

100 mL of agarose was suspended in 900 mL of distilled water containing 9 mmols of sodium periodate in order to ensure full oxidation of glyceryl groups. After 2 hrs, all glyceryl groups were completely oxidized and then the support was washed with an excess of distilled water.

Protocol 2   Preparation of aldehyde dextran

A 100 mL solution containing 3.33 g of dextran (MW 6,000 D) in distilled water was prepared. Then, 8 g of solid sodium periodate was added (this permitted the full oxidation of the dextran molecule) and this solution was stirred for 3 hrs. After this, the solution was dialyzed 4 times against 50 volumes of distilled water to eliminate the formaldehyde produced during the oxidation (around 180 mM) [12].

Protocol 3 Immobilization of PLE on glutaraldehyde-agarose

Enzymatic assay

The activity of this enzyme was followed by recording the increase in absorbance promoted by the hydrolysis of p-nitrophenyl propionate (pNPP) at 348 nm. The reaction mixture (0.4 mM pNPP in 25 mM sodium phosphate buffer pH 7.0) was stored in an ice-bath in order to reduce spontaneous hydrolysis. 2.5 mL of the reaction mixture was delivered into a 3 mL spectrophotometric cell and pre-heated at 25 °C for 15 min inside a spectrophotometer (Uvikon 930, Kontron Instruments) provided with thermostatic control and a stirring device. The reaction was started by adding the enzyme solution/suspension and the increase in absorbance per unit time was recorded from the linear portion of the curve. Extinction coefficient of released p-nitrophenol under these conditions was 5,150 $M^{-1}cm^{-1}$. One enzyme activity unit was defined as the amount of enzyme that hydrolyzes 1 micromole of substrate per min under the assay conditions.

Immobilization of PLE on glutaraldehyde supports

A 90 mL solution containing 1 mg of PLE in 0.1 M sodium phosphate, pH 7 and at 4 °C was prepared. Then, 10 mL of glutaraldehyde-agarose were added and the suspension was very gently stirred. A parallel experiment was performed adding inert agarose to check the stability of the enzyme and the absence of non-specific adsorption of the enzyme on the support. Periodically, samples of supernatant and suspension were withdrawn after centrifugation and their residual activity was measured as described above.

At the desired times, the enzyme support multi-interaction was ended by reduction of remaining aldehyde groups and aldehyde-amine bonds with sodium borohydride. The pH value was adjusted at 10 and 100 mg of solid sodium borohydride was added. After 30 min, the immobilized enzyme was filtered (e.g. using a Buchner flask with glass-sintered funnel connected to a vacuum line) and washed 3 times with 3 volumes of 0.1 M sodium phosphate pH 7 and then with an excess of distilled water.

Protocol 4 Immobilization of PLE on glyoxyl-agarose

A 90 mL solution containing 1 mg of PLE in 0.1 M sodium bicarbonate at pH 10 and 4 °C was prepared. Then, 10 mL of glyoxyl-agarose was added and the suspension was very gently stirred. Again, a reference suspension was prepared, using the inert agarose instead of the glyoxyl group. Periodically, samples of supernatant and suspension were withdrawn and their residual activity measured as described above.

At the desired times, the enzyme-support multi-interaction was stopped by reduction with sodium borohydride by adding 100 mg of solid sodium borohydride. After 30 min, the immobilized enzyme was filtered (as described above for the supports) and washed 3 times with 3 volumes of 0.1 M sodium phosphate pH 7 and an excess of water.

Protocol 5   Immobilization of poly (ethylenimine) on PLE-agarose

A solution of 10 mg/mL polyethyleneimine (PEI) (MW 25,000) was made in distilled water and the pH adjusted to 10.05 with conc. HCl in a cold bath. Then, a derivative was prepared following Protocol 4, but 30 min before reduction with sodium borohydride, 8 mL of PEI was added to the immobilization suspension. After 30 min, the derivative was reduced as described in Protocol 4.

Protocol 6   Chemical modification of enzyme/polyamine derivative with aldehyde dextran

10 mL of PLE/PEI co-immobilized derivative was suspended in 60 mL of sodium phosphate at pH 7 and 4 °C. Then, 30 mL of aldehyde dextran prepared as previously described was added to the suspension. This was very gently stirred for 24 hrs. Then, 900 mL of 0.1 M sodium borate, pH 8.5 and at 4 °C containing 2 g of sodium borohydride was added to reduce the remaining aldehyde groups as well as the aldehyde-amine bonds. After 1 hr, the enzyme derivative was washed 4 times with 3 volumes of 0.1 M phosphate and then with an excess of distilled water.

Protocol 7   Inactivation of PLE derivatives by organic solvents

The desired cosolvent-aqueous buffer solution was prepared and cooled in an ice bath. The pH was adjusted and no further corrections were performed. Then, the derivatives were equilibrated with this cooled solutions and 1 mL of PLE derivative was resuspended in 9 mL of the cosolvent/aqueous buffer mixture and incubated at the desired temperature. Periodically, samples of the suspension were withdrawn and its residual activity measured as described above.

4   Troubleshooting

The development of strategies of enzyme engineering via immobilization and post-immobilization techniques may involve complex concepts on reactivity between proteins and solid structures but finally the experimental protocols derived from such concepts are very easy to be carried out. Now, one small troubleshooting directly related with these specific experiments will be mentioned and two possible sequential solutions will be proposed:

4.1   Multipoint immobilization of proteins on glyoxyl agarose at pH 10.0

In order to force multipoint immobilization the use of slightly alkaline conditions is strictly necessary. Under such conditions soluble enzymes may be quite unstable. In order to solve such problems we would recommend two combined approaches:

- *Alkaline stabilization of soluble enzymes during immobilization can be intended by using soluble additives, e.g. polyols (like sorbitol, dextrans, poly-ethylene-glycol, glycerol and so on) [13]:*
In addition to that, salts and substrates or inhibitors of the enzyme have also been reported as very suitable stabilizing agent [14]. Obviously, additives, buffers, substrates and inhibitors should not contain amino groups in order to avoid competition with the enzyme during covalent immobilization.

- *Enzyme immobilization at a gradient of temperatures*
In some cases additives may be not enough to get a fairly stable soluble enzyme (e.g. preserving more than 75% of activity for 3 hrs at pH 10, 25 °C). In these cases, we can also perform a multipoint immobilization by using a gradient of temperatures under the best conditions (the most suitable additives). Initially, immobilization can be performed at low temperature (e.g. 4 °C) in order to improve enzyme stability. In this way, the degree of multipoint immobilization will be much lower than one reached at 25 °C [5] but it may be high enough to allow the immobilized enzyme to be further incubated at 15 °C for 1 hr in order to increase the intensity of multipoint immobilization. Finally, partially stabilized immobilized enzyme can be incubated at 25 °C for an additional 2 hrs period. In this way, we would have performed a very mild multipoint immobilization that could be so intense as a direct immobilization for 3 hrs at room temperature.

5   Remarks and Conclusions

5.1   Stability of multipoint immobilized derivatives of PLE in organic media

PLE immobilized on very poorly activated glutaraldehyde-agarose gels loses 50% of activity after 5 min in 50% acetonitrile at pH 7.0 and 25 °C. However, PLE immobilized on very highly activated glyoxyl-agarose preserves more than 90% of activity after 6 hrs under the same experimental conditions (Fig. 6). We had previously demonstrated that this latter immobilization procedure promoted a very intense multipoint covalent attachment of other industrial enzymes (e.g. trypsin [5] or penicillin G acylase [6]). In fact, such multipoint immobilized derivatives of trypsin and penicillin G acylase are much more stable

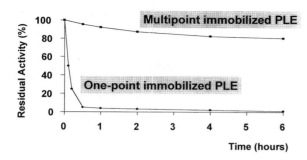

Figure 6 Time courses of inactivation of different PLE derivatives in the presence of organic cosolvents.

Both one-point and multipoint immobilized derivatives were incubated at 25 °C in 25 mM phosphate buffer pH 7.0 containing 50% of acetonitrile. At different times, aliquots of the suspension were withdrawn and assayed according the standard method (see Protocols).

than soluble enzymes and one-point immobilized derivatives against any type of denaturing agent (not only against organic cosolvents). In this way, we may assume that rigidification of enzyme molecules by multipoint immobilization on very highly activated supports (having more than 40 aldehyde residues below each immobilized molecule) may also be mainly responsible for this great increase of stability against organic solvents. Obviously the use of agarose, as a highly hydrophilic support, could also have some additional profitable effect.

5.2   Effect of different cosolvents on stability of PLE derivatives

In Figure 7 we observe very different time-courses of inactivation of multipoint immobilized PLE derivatives when incubated in the same concentration (60%) of three organic cosolvents: acetonitrile, dioxane and cellosolve. Acetonitrile is the most deleterious cosolvent. In fact, activity of the derivatives decreased

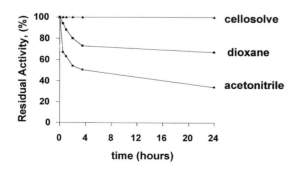

Figure 7 Time courses of inactivation of multipoint immobilized PLE derivatives in the presence of different organic cosolvents.

Multipoint immobilized derivatives were incubated at 25 °C in 25 mM phosphate buffer pH 7.0 containing 60% of organic cosolvent (acetonitrile, dioxane or cellosolve). At different times, aliquots of the suspensions were withdrawn and assayed according the standard method (see Protocols).

down to 50% after 4 hrs in the presence of this cosolvent. On the contrary, derivatives preserved 100% of activity after 24 hrs of incubation of a similar solution of cellosolve, the less harmful cosolvent. The mild effect of cellosolve, diglyme, tetraglyme and similar cosolvents had been previously observed with other enzymes (e.g. penicillin G acylases from *E.coli* and from *K. citrophila*) [15]. However in those cases, dioxane was much more deleterious than acetonitrile. It seems that some very general guidelines may be possible about the role of cosolvents on enzyme stability but each enzyme has to be experimentally investigated to obtain the qualitative picture. The very different behaviour of enzyme derivatives in the presence of different organic solvents means that it is possible to search for the best solvents to improve greatly the performance of enzyme catalyzed biotransformations. Several solvents having similar physico-chemical properties (e.g. ability to dissolve a given substrate, ability to shift thermodynamical equilibrium, etc.) may have a very different effect on activity-stability of enzyme derivatives.

5.3   Effect of cosolvent concentration

In addition to the nature of the solvent, cosolvent concentration plays a dramatic effect on stability of multipoint immobilized PLE derivatives (Fig. 8). A dramatic decrease in enzyme stability when increasing cosolvent concentration from 60 up to 80% is observed. Obviously, such apparently mild increase in cosolvent concentration is also representing a more acute decreasing in water concentration (from 40 to 20%). This decrease in water concentration could be mainly responsible for the high increase in enzyme instability. However, such decrease in water concentration may also exert some profitable effects in biotransformations (e.g. it seems to be necessary to get dramatic shifts in reaction equilibria [16]). In this way, the development of suitable techniques to improve the performance of enzyme under such harmful conditions is quite interesting in enzymatic organic synthesis.

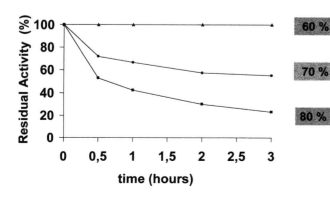

Figure 8   Time courses of inactivation of multipoint immobilized PLE derivatives in the presence of different concentrations of cellosolve cosolvents.

Multipoint immobilized derivatives were incubated at 25 °C in 25 mM phosphate buffer pH 7.0 containing 60, 70 or 80% of cellosolve. At different times, aliquots of the suspensions were withdrawn and assayed according the standard method (see Protocols).

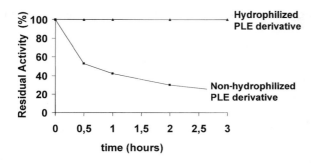

Figure 9 Time courses of inactivation of hydrophilized and non-hydrophilized PLE derivatives in the presence of high concentrations of organic cosolvents.

Both hydrophilized and non-hydrophilized multipoint immobilized derivatives were incubated at 25 °C in 25 mM phosphate buffer pH 7.0 containing 80% of cellosolve. At different times, aliquots of the suspension were withdrawn and assayed according the standard method (see Protocols).

5.4   Stability of fully hydrophilized immobilized derivatives

The sequential modification of PLE derivatives by co-immobilization of polyethyleneimine plus additional chemical modification with aldehyde-dextrans provides an additional profitable effect on their stability in the presence of high concentrations of organic solvents (Fig. 9). Hydrophilized and rigidified PLE derivatives retained full activity after 3 hrs of incubation in the presence of 80% of cellosolve but non-hydrophilized rigidified derivatives retain only 20% of activity after an identical incubation. As we commented in the introduction, the complete hydrophilization of immobilized enzymes may promote very different and diverse beneficial effects. However, keeping in view the results discussed, we may assume that a mild partition effect on the cosolvent promoted by such a hydrophilic nano-environment could perfectly explain the dramatic stabilizing effect of hydrophilization. The simple reduction of cosolvent concentration from 80% in the bulk solution down to 60% in the surroundings of the enzyme could be sufficient to promote such a dramatic increase in the stability of immobilized derivatives after hydrophilization.

The stability of enzyme derivatives and consequently the laboratory and industrial performance of enzyme-catalyzed biotransformations in organic media are greatly improved via the adequate and sequential combination of i) multipoint enzyme immobilization; ii) co-immobilization of hydrophilic polyamines; iii) chemical modification with aldehyde dextrans (which are converted into polyols after borohydride reduction); iv) a suitable selection of organic cosolvent. Exemplified by the stability of Pig Liver Esterase (PLE) in water-cosolvent mixtures we have demonstrated the high efficiency of such a stabilization strategy. For example, we have been able to prepare immobilized derivatives that retained full activity after 3 hrs at 25 °C in 80% of cellosolve as cosolvent.

On the other hand, non-stabilized derivatives retained less than 50% of activity after incubation in only 60% acetonitrile for only 1 min. In this way the operational stability of the biocatalyst may be increased by several orders of magnitude.

Acknowledgments

The authors wish to thank R. Armisen for their stimulating discussions and Hispanagar S.A. for the gift of agarose gels. We are also very grateful to M.C. Ceinos for her skilled technical assistance. This work has been supported by the European Community (BIO4-CT96–0005).

References

1   Kasche V, Haufler U, Riechman L (1987) Equilibrium and kinetically controlled synthesis with enzymes: semisynthesis of penicillins and peptides. *Methods in enzymol* 136: 280–292

2   Bes T, Gómez-Moreno, Guisán JM, Fernández-Lafuente R (1995) Selective enzymic oxidations: Stabilisation by multipoint covalent attachment of ferredoxin NADP-reductase, an interesting cofactor recycling enzyme. *J Mol Cat* 98: 161–169

3   Bastida A, Sabuquillo P, Armisen P et al. (1988) A single step purification, immobilization and hyperactivation of lipases via interfacial adsorption on strongly hydrophobic supports. *Biotechnol Bioeng* 58: 48–493

4   Guisan JM (1988) Aldehyde gels for immobilization-stabilization of enzymes. *Enzyme Microb Technol* 10: 375–382

5   Blanco RM, Guisan JM (1989) Stabilization of enzymes by multipoint covalent attachment to agarose-aldehyde gels. Borohydride reduction of trypsin-agarose derivatives. *Enzyme Microb Technol* 11: 360–366

6   Fernandez-Lafuente R, Rosell CM, Guisan JM (1991) Enzyme reaction engineering: synthesis of antibiotics catalyzed by stabilized penicillin G acylase in the presence of organic solvents. *Enzyme Microb Technol* 13: 898–905

7   Guisàn JM, Bastida A, Blanco RM et al. (1997) Immobilization of enzymes on glyoxyl supports: strategies for enzyme stabilization by multipoint covalent attachment. In: Bickerstaff, G (ed.): *Immobilization of Enzymes and Cells* Serie Methods Biotechnol. Vol 1 The Humana Press Inc. NJ (E). 277–288

8   Guisán JM, Álvaro G, Fernández-Lafuente R et al. (1993) Stabilization of an heterodimeric enzyme by multipoint covalent immobilisation: Penicillin G acylase from *Kluyvera citrophila. Biotechnol Bioeng* 42: 455–464

9   Fernandez-Lafuente R, Rosell CM, Caanan-Haden L et al. Facile synthesis of artificial enzyme nano-environments via solid-phase chemistry of immobilized derivatives: dramatic stabilization of penicillin G acylase versus organic solvents. *Enzyme Microb Technol;* 84: 96–103

10 Wong CH, Whitesides GM (1994) Enzymes in synthetic organic chemistry. Tetrahedron Organic Chemistry series Volume 12. Pergaman, Oxford

11 Fernández-Lafuente R, Rosell CM, Rodríguez V et al. (1993) Preparation of activated supports containing low pK amino groups. A new tool for protein immobilisation via the carboxyl coupling method. *Enzyme Microb Technol* 15: 556–550

12 Guisan JM, Rodriguez V, Rosell CM et al. (1997) Stabilization of immobilized enzymes by chemical modification with polyfunctional macromolecules. In: *Methods in Biotechnology 1. Immobilization of enzymes and Cells* (Bickerstaff, G.F. ed) Humana press. Totowa, New Jersey

13 Kazan D, Erarslan A (1996) Influence of polyhydric compounds on the pH stability of penicillin G acylase obtained from a mutant of Escherichia coli ATCC 11105. *Process Biochem* 31: 691–97

14 Álvaro G, Fernández-Lafuente R, Blanco RM, Guisán JM (1991) Stabilizing effect of penicillin G sulfoxide, a competitive inhibitor of penicillin G acylase: its practical applications. *Enzyme Microb Technol* 13: 1–5

15 Rosell CM, Terreni M, Fernández-Lafuente R, Guisán JM (1998) A criterium for the selection of monophasic solvents for enzymatic synthesis. *Enzyme Microb Technol;* 83: 64–69

16 Fernandez-Lafuente R, Rosell CM, Guisan JM (1996) Dynamic reaction design of enzymic biotransfomations in organic media: equilibrium controlled synthesis of antibiotics by penicillin G acylase. *Biotechnol Appl Biochem* 24: 139–143

Immobilization of Lipases for Use in Non-Aqueous Reaction Systems

John A. Bosley and Alan D. Peilow

Contents

1   Introduction

Although the natural function of lipases is to hydrolyse fats and oils it has been shown that, under very low water conditions, the reverse reaction dominates and the lipases can be utilised for the catalysis of esterification and transesterification reactions. Although crude suspensions of enzyme powders have successfully been used for these reactions [1] their effectiveness (efficiency) is, in general, much lower than for immobilized lipase biocatalysts. In order to fully

Methods and Tools in Biosciences and Medicine
Methods in non-aqueous enzymology, ed. by M. N. Gupta
© 2000 Birkhäuser Verlag Basel/Switzerland

exploit lipases under these essentially nonaqueous (microaqueous) conditions the use of immobilization technology offers a number of important benefits, including; enzyme re-use, easy separation of product from enzyme and the potential to run continuous processes via packed-bed reactors. In some cases the activity and stability of the enzyme is also improved [2].

There is now a great deal of literature describing the many methods for the immobilization of lipases, much of which has been collated in reviews [3, 4]. These methods can be split into the following broad groups; drying, precipitation or adsorption on hydrophilic particles such as silica and alumina [5, 6]; adsorption on hydrophobic surfaces [2, 7–9]; adsorption on ion-exchange resins [10–12]; covalent attachment to suitable ligands [13–15] and physical entrapment in polymer [16] or hydrophobic silica gels [17]. It has also been reported that cross-linked enzyme crystals (CLEC) are effective biocatalysts in nonaqueous media [18] but as this method requires relatively pure lipase it will not be considered further in this chapter.

It is also important to choose a support material that allows the immobilized lipase to perform at maximum efficiency [19]. To function effectively a potential support material must satisfy a number of important criteria; it must allow easy enzyme immobilization without appreciable losses of activity; it should have appropriate pore and particle size so as not to limit diffusion of substrates and therefore reaction rates [6, 20]; it should have suitable mechanical properties for the desired process; it must contain no extractable materials likely to contaminate the product stream [10]; and ideally it should be inexpensive.

The most widely used methods are those based on adsorption processes. These are generally simple, effective, do not require the use of potentially harmful chemicals and do not result in large losses of lipase activity. Consequently the protocols described in this chapter concentrate on these systems. The use of covalent binding or polymer entrapment will not be covered and the reader should refer to the references listed above for further details of these methods.

2   Materials

Chemicals

Ethyl palmitate	Aldrich
Lauric acid	Fluka Chemica
Methyltrimethoxy silane (MTMS)	Fluka Chemica
Octadecyltrimethoxysilane (ODTMS)	Aldrich
Octan-1-ol	Fisher
Oleic acid	BDH
Ovalbumin (grade V)	Sigma

Polydimethylsiloxane (PDMS,
   silanol terminated, MW 550)   ABCR GmbH
Propyltrimethoxysilane (PTMS)   Aldrich
Tetramethoxysilane (TMOS)   Aldrich
All other chemicals were standard laboratory grade.

Enzymes
- Lipase B: *Candida antarctica* (SP434, 180,000 LU/g), gift from Novo-Nordisk (Copenhagen, Denmark).
- Lipase N: *Rhizopus niveus* (3,200 LU/g) was obtained from Amano Enzymes (Nagoya, Japan).
- Lipase SP388: *Rhizomucor miehei* gene-cloned into *Aspergillus oryzae* (100,000 LU/g) gift from Novo-Nordisk (Copenhagen, Denmark).
- Lipase Type VII: *Candida rugosa* (8,500 LU/g) was obtained from Sigma-Aldrich Co. Ltd (Poole, UK).
- Lipolase: *Humicola sp* (100L, 100,000 LU/mL) gift from Novo-Nordisk (Copenhagen, Denmark).
- Lipozyme: *Rhizomucor miehei* (10,000L, 10,500 LU/mL) gift from Novo-Nordisk (Copenhagen, Denmark).

All lipase preparations were used as supplied, without further purification.

Equipment
- Support: Accurel® EP100 (macroporous polypropylene) was obtained from Akzo Nobel (Obernburg, Germany).
- Support: Celite® R648 was obtained from Manville Corporation (Twickenham, UK).
- Support: Duolite® ES568N (macroporous anion exchange resin) was obtained from Rohm and Haas (Croydon,UK).

3   Methods

3.1   Tributyrin hydrolysis assay

The activity of lipase solutions is determined using an assay based on a standard Novo Nordisk assay [21]. A sample of lipase solution (typically up to 5 mL containing a total of 1–10 Lipase Units) is added to a gum arabic-stabilized suspension of tributyrin in distilled water (20 mL) at pH 7 and at 30°C. The suspension is prepared by homogenising a mixture of tributyrin (3 mL), distilled water (47 mL) and emulsification reagent (10 mL) at 24,000 rev. min^{-1} for 15 secs (Ultra -Turrax Model T25, Janke & Kunkel GmbH, Staufen, Germany). The emulsification reagent is prepared by dissolving gum arabic (6.0 g), glycerol (540 mL), NaCl (17.9 g) and KH_2PO_4 (0.41 g) in distilled water (400 mL) and then making up the volume to 1000 mL with distilled water. After addition of the lipase, the pH is maintained at pH 7 in an autotitrator (VIT 90 Autotitrator, ex Radiometer Copenhagen, Denmark) by titration with 100 mM NaOH solution using a 1 mL burette. One lipase unit (LU) is defined as the amount of lipase that liberates 1 µmol butyric acid per minute under these conditions. Lipase loadings are expressed as thousands of lipase units per gram of dry support (kLU/g). For Protocols 1 and 2, which are based on passive adsorption, the lipase loading is defined as the amount of activity lost from solution. For Protocols 3 and 4, which are based on drying or entrapment, the lipase loading is simply the number of lipase units added.

3.2 Esterification assay (octyl oleate)

The immobilized lipase (typically 5–20 mg of the vacuum dried material, depending on lipase loading) is placed in a crimp-seal vial, and a mixture of oleic acid (5.88 g), octan-1-ol (2.70 g) and distilled water (0.175 g) is added. The vial is sealed and placed in a waterbath at 50°C for 30 min, and shaken at 200 strokes/min. A sample (100 µL) is then removed and immediately eluted down a small alumina column (2 mL, basic, activity II) with diethyl ether, together with a solution of methyl stearate (2.50 mg) dissolved in petroleum ether (100 µL, b.p. 100–120°C) as an internal standard. The diethyl ether is then removed by evaporation and replaced with petroleum ether (4 mL, b.p. 60–80°C). The ratio of octyl oleate to methyl stearate is then determined by Gas-Liquid chromatography – (for GC column and conditions see Appendix). The activity is expressed as µmol ester formed per min per mg catalyst. The lipase efficiency is obtained by dividing lipase activity by lipase loading.

3.3 Esterification assay (octyl laurate)

As in the octyl oleate assay except that each vial contains lauric acid (5.30 g), octan-1-ol (3.52 g) and water (0.105 g). Ethyl palmitate is used as the internal standard.

3.4 Protocols

Protocol 1   Adsorption on hydrophobic surfaces

1. Wet a sample of Accurel® EP100 (2.0 g) by adding absolute ethanol (6 mL) and then add 10 mM phosphate (NaH_2PO_4) pH 7 buffer (54 mL). Mix the suspension gently on a orbital shaker for 15 min.

2. Dissolve ovalbumin (0.26 g) in the same buffer (90 mL) and add to the Accurel® EP100 suspension. Continue gentle mixing for 1 hr.

3. Dissolve the required amount of lipase in the same buffer (100 mL) and add to the wetted Accurel® EP100. Leave the sample mixing on an orbital shaker at room temperature (approximately 20°C).

4. Monitor the adsorption of lipase onto the support by measuring loss of activity from the solution using the tributyrin hydrolysis assay (see 3.1.). Remove a sample (0.50 mL) at about hourly intervals and determine its lipase activity. The adsorption normally takes 4–16 hrs to reach completion (> 90% of activity adsorbed), depending on the lipase preparation and the required loading.

5. Collect the catalyst by vacuum filtration using filter paper (Whatman No. 1), wash twice in 10 mM phosphate pH 7 buffer (100 mL), and rinse with distilled water several times to remove buffer salts and unadsorbed material. Leave the sample under suction on the filter paper for 1 hr to remove the majority of the water.

6. Complete drying by placing the sample in a vacuum oven at room temperature until no further weight loss is recorded (normally overnight).

Protocol 2   Adsorption on ion-exchange resins

1. Determine water content of Duolite® ES568N by measuring the weight loss of a sample after drying for 16 hrs at 110°C (the water content of Duolite can vary and lipase loading is expressed on a dry weight basis).

2. Place sample of Duolite® ES568N (equivalent 1.00 g dry weight) in 100 mM phosphate pH 6.0 buffer (100 mL) and mix gently for 1 hr. Transfer ion-exchange resin to fresh buffer and leave for a further 1 hr. Transfer ion-exchange resin to fresh 100 mM phosphate pH 6.0 buffer (100 mL).

3. Add to the Duolite® ES568N suspension the required amount of lipase dissolved in phosphate pH 6.0 buffer (100 mL). Leave the Duolite® suspension mixing gently on an orbital shaker.

4. Monitor the adsorption of lipase onto the support by measuring loss of activity from the solution using the tributyrin hydrolysis assay. A sample (0.50 ml) is removed at about hourly intervals and its lipase activity determined. The adsorption normally takes 4–16 hrs to reach completion (90% of activity adsorbed), depending on the lipase preparation and the required loading.

5. Collect the catalyst by vacuum filtration using filter paper (Whatman No. 1), wash twice in 100 mM phosphate pH 6 buffer (100 mL), and rinse several times with distilled water to remove unadsorbed material. Leave the sample under suction on the filter paper for 1 hr to remove the majority of the water.

6. Complete drying by placing the sample in a vacuum oven at room temperature until no further weight loss is recorded (normally overnight).

Protocol 3   Precipitation/drying onto porous silica (Celite®)

1. Dissolve the required amount of lipase in water (5 mL).

2. Add Celite® R648 (10.0 g) to the solution and mix to ensure the particles are evenly wetted.

3. Dry the particles at room temperature under vacuum for 20 hrs.

   Note:
   If other grades of Celite®, e.g. Hyflo-Supercel™, are used it may be necessary to break up any lumps formed during the drying process.

Protocol 4   Physical entrapment in hydrophobic sol-gel materials

1. Dissolve required amount of lipase in distilled or de-ionized water (400 µL) and add to a mixture containing water (164 µL), 1 M sodium fluoride solution (100 µL), and 4% w/w polyvinyl alcohol solution (MW 15,000, 200 µL).

2. Add tetramethoxy silane (150 µL) and octadecyltrimethoxysilane (1950 µL) to the mixture, and mix vigorously for 5 sec on a vortex mixer and then gently shake by hand. The mixture should form a clear homogenous solution after 30 sec.

3. Cool the mixture for 5 min in an ice-water bath (~ 0°C) to prevent loss of activity due to the exothermic reaction. The sample is then left for 24 hrs at ambient temperature to allow the reaction to go to completion.

4. Dry the resultant gel for 3 days at 37°C.

5. Break up the dry gel into a powder using a mortar and pestle and incubate for 2 hrs in water (10 mL).

6. Collect the gel by filtration, wash in water, acetone and pentane, and dry overnight at 37°C.

4   Troubleshooting

- *In order to measure lipase loading accurately it is essential that the lipase solution is stable over the required period. Apparent loss of activity can arise from unexpected pH changes (e.g. caused by residues from the support), precipitation (as a result of pH changes or addition of some ionic species), or proteolysis (some lipase preparations contain significant protease activity).* If possible, supports should be washed prior to use. If proteolysis is thought to be a problem it can often be minimised by lowering the immobilization temperature (e.g. to ~5°C).

- *Some lipase preparations, particularly powdered forms, can contain significant levels of insoluble material.* This should be removed by filtration or centrifugation before immobilization as this material can lead to pore blockage and results in poor immobilization and/or low catalyst activity.

- *Care should be taken to ensure the support is fully wetted both before the lipase solution is added and during the immobilization process.* This can be difficult with some materials such as polypropylene that tend to float and can form a dry crust on the enzyme solution. A small amount of mechanical agitation should be used to maintain good contact between enzyme solution and support. The use of magnetic stirrer bars should be avoided for rigid support particles (e.g. silicas, some polymers) as excessive particle attrition can occur.

5   Remarks and Conclusions

5.1   Measurement of immobilized lipase activity (hydrolysis vs esterification)

Activity measurement for a lipase solution is normally carried out using an assay system based on the ability of the lipase to hydrolyse a suspension of triacylglyceride in water. Application of this type of assay to immobilized lipase usually leads to low apparent activity due to problems of mass transfer of substrate into the support used for immobilization. An example of this effect is seen when *Rhizomucor miehei* lipase is immobilized onto Accurel® EP100 (porous polypropylene). The hydrolysis activity of the catalysts was measured using the tributyrin assay and it was found that efficiency (activity divided by loading) fell rapidly as loading increased [8]. This results from changes in the distribution of lipase in the support as a function of loading. Increasing the lipase loading increases the penetration of lipase into the interior of the support particles. The substrate cannot access the interior due to mass transfer limitations and therefore lipase in the interior of the particles are inactive, and efficiency decreases. It may be expected that these mass transfer problems are particularly severe in this system as the substrate is in the form of a suspension of small droplets in water (~ 10 μm), most of which are larger than the mean pore diameter of the support (~ 0.2 μm). The use of heterogeneous substrate mixtures is therefore generally unsuitable for the assay of enzymes immobilized in porous supports and this might account for the relatively low (< 20%) recovered activity often reported. To allow comparison between different immobilization methods the preferred assay of immobilized lipase activity is to measure the initial rates of esterification using a single phase system of an equimolar mixture of fatty acid and alcohol containing a small amount of water dissolved in the reaction medium.

5.2 Hydrophilic vs hydrophobic supports

The natural role of lipases is to hydrolyse oils and fats and consequently they have both hydrophobic and hydrophilic patches on their surface to enable them to occupy the oil-water interface. The hydrophobic patches tend to be concentrated in the region of the lid covering the active site whereas most of the hydrophilic areas are on the opposite side of the molecule. Either of these regions can be effectively utilised for lipase immobilization. For adsorption on ionic hydrophilic supports the lipase is immobilized via ionic interaction. At neutral pH most lipases carry an overall negative charge (their isoelectric point, pI, is usually in the range 4–6) and therefore will be adsorbed on positively charged surfaces. Weak base anion exchange resins are ideal for this application, the use of strong base anion exchange resins can lead to enzyme deactivation. Alternatively lipases will readily adsorb on hydrophobic surfaces via van der Waals forces. In general, lipases do not adsorb on uncharged hydrophilic surfaces (such as Celite) and these materials only act as a physical support; there is no specific interaction with the lipase.

5.3 Porous hydrophobic supports

Previous work has established the ideal physical and surface chemistry requirements of support materials for the immobilization of lipases via hydrophobic interactions [20]. However the high cost of the controlled pore glasses used in that study prevents them from being used in most commercial applications. Supports based on macroporous polypropylene (Accurel® EP100) are now becoming widely used for lipase immobilization [8, 9, 22–24] as they largely fulfil the key physical requirements and are relatively inexpensive. The immobilization procedure is a simple mixing, filtering and drying process. The hydrophobic polymer is first wetted with a water miscible solvent, such as ethanol, the lipase solution added and the adsorption of the lipase allowed to come to equilibrium. The catalyst is then collected by filtration, washed and dried (see Protocol 1 for more details). The adsorption process usually takes between 4 and 16 hrs to complete depending on the amount of lipase being adsorbed. Examples of adsorption profiles for *C. rugosa* and *R. niveus* lipase are shown [Fig. 1]. It can be seen that the adsorption process is fast, over 80% of the *C. Rugosa* added was immobilized in less than 30 min.

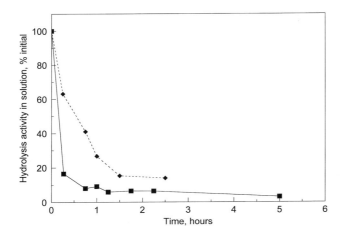

Figure 1 Adsorption profiles for *Rhizopus niveus* lipase (◆) and *C. rugosa* lipase (■) immobilized onto Accurel® EP100.

The amount of activity remaining in solution is expressed as a percentage of the initial activity.

5.4 Activity vs lipase loading

By varying the amount of lipase used in the immobilization a wide range of lipase loadings has been studied for several lipases [8]. It is interesting to note that high loadings of many lipases can be achieved, despite many of the commercial preparations containing high levels of other proteins. This infers that lipase is preferentially adsorbed on this polypropylene support. The effect of lipase loading on esterification activity for selected lipases is shown in Figure 2. As expected, the activity of the catalyst increases as the lipase loading is increased. At very high lipase loadings the effect of mass transfer limitations is seen as the activity versus loading relationship reaches a plateau.

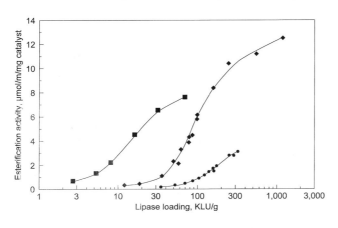

Figure 2 Effect of lipase loading on esterification activity for *Rhizomucor miehei* lipase (◆), *Humicola* sp. lipase (•), and *Candida antarctica* B lipase (■), immobilized onto Accurel® EP100.

As is shown, the plot at low loading is not linear; activity appears to be suppressed giving the plot an S shape. This effect is more obvious when the activity data is plotted in the form of efficiency (activity/loading) against loading. The efficiency of activity for *Rhizomucor miehei*/Accurel® EP100 catalysts is shown in Figure 3. The plot shows that at low loading, efficiency rapidly increases as loading is increased until it reaches a maximum at a loading of approximately 100 kLU/g. Thereafter efficiency decreases with increasing loading due to mass transfer limitations. The suppression of activity at low loading is possibly the result of conformational changes in the enzyme molecule. At low loading there is a large excess of surface available for adsorption. Lipase has a strong affinity for the surface (this is the basis of the immobilization method) and at low loading the enzyme attempts to maximise its contact with the surface. This results in a loss of conformation and consequently a reduction in efficiency. As loading is increased, the area available are for the lipase to adsorb decreases and therefore the lipase is less able to spread itself, leading to an increase in the amount of active conformation retained, and the efficiency increases. An alternative explanation is that the surface contains areas of high affinity (hot-spots) that lead to conformational change on adsorption and these areas adsorb lipase preferentially.

It is interesting to note that these effects cannot be seen in the hydrolysis data [8] due to dominance of the mass transfer limitations imposed on the heterogeneous substrate by the porous structure of Accurel® EP100 (see earlier).

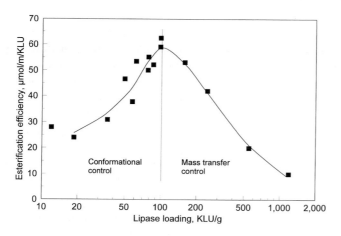

Figure 3 Effect of lipase loading on esterification efficiency for *Rhizomucor miehei* lipase immobilized onto Accurel® EP100.

5.5 Effect of protein treatment on Accurel® EP100 catalysts

The depression of activity at low loading can be overcome by treating Accurel® EP100 with a non-lipase protein prior to the adsorption of lipase. The exact mechanism by which the addition of other proteins improves lipase activity is not clear. It is possible that the proteins occupy sites of high affinity on the support or they may simply reduce the excess surface area available and therefore inhibit lipase deactivation. A number of proteins have been shown to be effective in stopping the deactivation at low loading for a range of lipases [25]. A plot of efficiency of activity versus loading for *Humicola* sp. lipase immobilized onto Accurel® EP100 treated with ovalbumin is shown in Figure 4. At low loading the deactivation process is inhibited and a large increase in efficiency over the untreated sample is observed. This suggests *Humicola* sp. lipase is prone to unfolding on adsorption to hydrophobic surfaces in the absence of protein treatment.

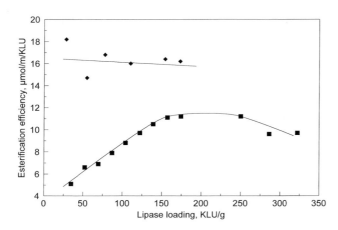

Figure 4   Effect of loading on the esterification efficiency of *Humicola sp* lipase on Accurel® EP100 (■) and *Humicola* sp lipase on Ovalbumin treated Accurel® EP100 (♦).

Not all lipases suffer from loss of activity at low loading. The B lipase from *C. antarctica* does not show increased activity when adsorbed onto ovalbumin treated Accurel® EP100 [Fig. 5]. The plot of efficiency versus loading shows a plateau at low loading and thereafter a decrease in efficiency with increasing loading as the mass transfer limitations are encountered. This infers that the structure of *C. antarctica* B is less prone to unfolding as a result the adsorption process, this could be the result of it having a weaker affinity for the support surface or a more rigid conformational structure.

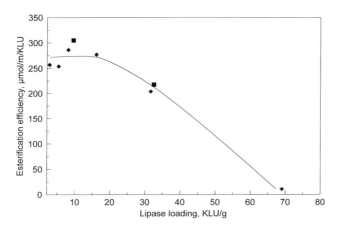

Figure 5   Effect of loading on the esterification efficiency of *C. antarctica B* lipase on Accurel® EP100 (♦) and *C. antarctica B* lipase on Ovalbumin treated Accurel® EP100 (■).

We have found the use of Accurel® EP100 as a support for lipase enzymes to be very simple and straightforward. Lipases are adsorbed readily and high loadings can be achieved. The resulting catalysts are highly active in esterification reactions. Some activity loss can be encountered at low loadings but this can be overcome by treatment of Accurel® EP100 with another protein (e.g. ovalbumin).

5.6 Ion-exchange resins

An alternative to adsorption via hydrophobic interactions is to use the ionic character of proteins to adsorb enzymes onto porous solids containing charged groups. In particular, a macroporous ion-exchange resin prepared by derivatising a phenol-formaldehyde polymer with diethylaminoethane (Duolite® ES568N) has been successfully used for the immobilization of *Rhizomucor miehei* lipase [10, 26]. Duolite® ES568N is weakly basic and therefore will be positively charged at neutral pH. Most lipases have isoelectric points (pI) below pH 7 and are negatively charged at this pH. Consequently they will be adsorbed via ionic interaction. As with the hydrophobic supports the immobilization process is a simple mixing and drying procedure. The ion-exchange resin is first adjusted to a neutral pH, the lipase solution added and the adsorption process allowed to go to completion. The resin is then separated, washed and dried (see Protocol 2). The lipase loading and activity for selected lipases adsorbed onto Duolite® ES568N are shown in Table 1.

Table 1 Lipase loading and esterification activity data for lipases adsorbed onto Duolite® ES568N.

Lipase Source	Saturation Loading [1] (kLU/g support)	Esterification Activity [2] (µmol/min/mg catalyst)
Rhizomucor miehei (native, Lipozyme)	9.9	0.46
Rhizomucor miehei (gene-cloned, SP388)	71.3	2.58
C. antarctica B	6.1	0.06

[1] Lipase loading (kLU) is the maximum amount adsorbed per g Duolite®.
[2] Esterification activities were measured using the octyl oleate esterification assay.

One disadvantage of using ion-exchange resins is that not all lipases can be immobilized effectively. In general, these supports have a lower capacity for lipases than the polypropylene support described in the previous protocol. This results from a number of factors of which the following are the most significant:

- many ion-exchange resins have small pores (e.g. Duolite ES568 has a mean pore diameter of ~ 50 nm, many others have pores of < 30 nm diameter) making it difficult for the lipase to utilize the internal surface area effectively;
- there is no discrimination between the lipase and other proteins present in the preparation.

The latter point is of great importance as many commercial lipase preparations contain high levels of other proteins. This is illustrated by the difference between the native *Rhizomucor miehei* (Lipozyme 10,000 L) and the much purer sample of the gene-cloned *Rhizomucor miehei* (SP388). It can be seen from Table 1 that much higher saturation loadings can be obtained for the more highly purified lipase preparation. *C. antarctica* B lipase is relatively poorly adsorbed on this carrier and the resultant catalyst has low activity compared to similar loadings using Protocols 1 or 3.

Other ion-exchange resins can also be used successfully to adsorb and obtain active lipase catalysts (e.g. Fractogel® TSK 650 M obtained from Merck, UK).

5.7   Precipitation / drying onto porous silica (Celite®)

Diatomaceous earth or kieselguhr, commonly known as Celite®, has been used as a lipase carrier for a number of years. Celite® is the skeletal structure of diatoms and occurs as fossil deposits in large abundance throughout the world. The main constituent of Celite® is silica and forms an inert porous structure whose properties give rise to its use as a filter aid. The large pore structure promotes lipase activity by dispersing the lipase over the surface of the pores increasing the substrate/lipase interfacial area. Catalysts are simply prepared by mixing Celite® with lipase solution and removal of the water by evaporation. Alternatively, lipase can be precipitated onto the support by addition of acetone to the solution. Lipase trapped in the Celite® structure is not tightly bound and is readily dissolved and removed from the support in the presence of excess water. Fully dried Celite® catalysts prepared by drying/precipitation may have low apparent activity, but can be activated by addition of water [27]. Replacement of water by glycerol, various diols, fatty acids, fatty acid esters, amino acids, peptides and phospholipids have also been claimed to be effective [28–30]. Some examples of Celite® immobilized lipases are shown in Table 2.

Table 2 Examples of the activity obtained by drying lipase onto porous silica (Celite® R-648).

Lipase	Loading (kLU/g)	Esterification activity [1] (μmol/min/mg)
Rhizomucor miehei (Lipozyme 10,000 L)	19.8	0.94
C. antarctica B	7.8	1.23

[1] Esterification activity determined using octyl laurate esterification assay.

Drying of lipase on Celite® is a very simple method of immobilization and has been used by a large number of workers. However, although it generates active catalysts, these are usually much less active than similarly loaded materials prepared by adsorption on hydrophobic supports. Immobilization on Celite® results in catalysts that are essentially lipase particles distributed throughout an inert carrier and as such are much closer to enzyme powders than the monolayers of enzyme molecules obtained with Protocols 1 and 2.

5.8   Sol-gel immobilization

The use of the sol-gel process to entrap lipases involves the hydrolysis of $Si(OR)_4$ via acid or base catalysis to form an amorphous network of SiO_2 to form around the enzyme. Reetz and co-workers have found that traditional silica gel processes involving $Si(OCH_3)_4$ led to low activities, but addition of a hydrophobic alkyl silane $CH_3Si(OCH_3)_3$ to the mixture led to large increases in activity [17]. Further work has shown that the choice of alkyl silane used in the process strongly influences the activity obtained [31]. The increase in activity on incorporation of hydrophobic silane reagents is believed to be due to a combination of increasing the transport of hydrophobic substrate into the gel and interfacial activation of the lipase. The protocol for the sol-gel process (Protocol 4) describes the use of $Si(OCH_3)_4$ and $C_{18}H_{37}Si(OCH_3)_3$ as the reagents. Examples of *Humicola* sp. and *C. antarctica B* lipases immobilized using a range of hydrophobic silane reagents are shown in Table 3.

Table 3 Lipase loading and activity of catalysts produced by entrapment of lipase in hydrophobic silica gels.

Lipase	Silane reagents [1] (molar ratio)	Lipase loading [2] (kLU/g gel)	Esterification activity [3] (µmol/min/mg)
Humicola sp.	MTMS	22.3	0.00
	MTMS/PDMS (6:1)	12.4	0.00
	TMOS/PrTMS (1:5)	21.4	0.09
C. antarctica B	MTMS	95.0	0.03
	MTMS/PDMS (6:1)	58.0	0.03
	TMOS/PrTMS (1:5)	61.0	0.77
	TMOS/BuTMS (1:5)	73.6	1.60
	TMOS/ODTMS (1:5)	24.6	1.89

[1] MTMS = $CH_3 Si(OCH_3)_3$, PDMS = $HO[(CH_3)_2Si-O]_nSi(CH_3)_2OH$, TMOS = $Si(OCH_3)_4$, PrTMS = $n-C_3H_7Si(OCH_3)_3$, BuTMS = $n-C_4H_9 Si(OCH_3)_3$, ODTMS = $n-C_{18}H_{37} Si(OCH_3)_3$.
[2] Lipase loading is the amount of activity added at the start of the sol-gel process.
[3] Esterification activity was determined using the octyl oleate esterification assay.

The effect of increasing activity by increasing the hydrophobicity of the silane gel has also been demonstrated for *C. antarctica* B lipase. However the activities obtained for both lipases are lower than that obtainable by using hydrophobic adsorption onto Accurel® EP100 [Fig. 3]. It is probable that the sol-gel system will suffer from some mass transfer limitations due to its pore gel structure in comparison to the more open structure of Accurel® EP100.

5.9   Effect of water activity (A_w) on esterification reaction rate

Although (trans)esterification reactions using immobilized lipase catalysts will take place under very dry conditions [32] the reaction rate can often be greatly increased by the presence of small amounts of water to maintain enzyme hydration. As all the components of the reaction mixture (solvents, substrates, products, support, enzyme) will be capable of binding water to varying degrees, the thermodynamic water activity (A_w), that is related to water content via the water adsorption isotherm [33, 34], is a more important parameter than water content in these essentially nonaqueous reaction systems [35]. The activity of the catalyst is also very dependant on the source of the lipase, for example, *Humicola* sp. lipase shows very low esterification activity at low water activity ($A_w = 0.1$) compared to intermediate values ($A_w = 0.6$–0.7). In contrast the esterification activity of *Rhizopus niveus* lipase at high and low water activities is similar [33].

A direct consequence of these water activity effects is that initial esterification reaction rates can be very slow if the reaction mixture is dry [36]. However, as water is a product of the reaction, the water activity will increase as a function of time and the reaction rate will tend to accelerate. This can result in a non-linear reaction profile that makes it difficult to measure accurately and reproducibly the esterification activity of the catalyst. The slow initial rate can be increased by the addition of a small amount of water (see Methods) to the reaction mixture leading to a linear reaction profile until high conversions are reached. Alternatively it has been shown that A_w can be maintained at a fixed value by use of appropriate salt hydrates [37, 38]. It should be noted that the addition of too much water may lead to reduced degrees of conversion as the competing hydrolysis reaction will become significant. In some cases excessive water levels may also lead to rapid catalyst deactivation, particularly at elevated temperatures.

In order to achieve high esterification conversions it may be necessary to remove the generated water from the reaction system. This can be achieved by a number of methods including; use of vacuum [39], gas bleeds, pervaporation [40], cold fingers, molecular sieves [36].

Other types of reaction, e.g. acidolysis or transesterification, will not generate water as a by-product. In these systems, achieving the optimal water activity may be more difficult. The water activities needed to maintain sufficient hydration of the lipase in order to achieve acceptable reaction rates may lead to significant levels of substrate/product hydrolysis.

Acknowledgments

We gratefully acknowledge Dr. Albin Zonta for the immobilization of *Humicola* sp. and *C. antarctica* lipases in the sol-gel materials. We would also like to thank Mr S. Trevallion and Mrs C. Austin for their technical assistance in the immobilization studies.

Appendix

GC analysis of fatty acid esters

A Hewlett-Packard HP5980 chromatograph fitted with an Flame ionization detector (FID) and a non-polar fused silica wide-bore column (BP1, length 25 m, film thickness 1 µm, 0.53 mm diameter obtained SGE Europe Ltd, Milton Keynes, UK) is used to separate the standards from the octyl fatty acid esters. Helium is used as the carrier gas (5 p.s.i.). The table below lists the GC conditions used for both octyl oleate and octyl laurate assays.

Condition	Octyl laurate assay	Octyl oleate assay
Injection volume (µL)	1	2
Inlet temperature (°C)	250	250
Detector temperature (°C)	330	350
Oven conditions:		
Initial temp (°C)	220	220
Ramp rate (°C/min)	15	10
Final temperature (°C)	320	300

References

1 Wu XY, Jääskeläinen S, Linko YY (1996) An investigation of crude lipases for hydrolysis, esterification, and transesterification. *Enzym Microb Technol* 19: 226–231

2 Pencreac'h G, Baratti JC (1997) Activity of *Pseudomonas cepacia* lipase in organic media is greatly enhanced after immobilization on a polypropylene support. *Appl Microbiol Biotechnol* 47: 630–635

3 Malcata FX, Reyes HR, Garcia HS et al. (1990) Immobilized lipase reactors for modification of fats and oils – A review. *JAOCS* 67(12): 890–910

4 Balcão VM, Paiva AL, Malcata FX (1996) Bioreactors with immobilised lipases: State of the art. *Enzym Microb Technol* 18: 392–416

5 Wisdom RA, Dunnill P, Lilly MD, Macrae AR (1984) Enzymic interesterification of fats: factors influencing the choice of support for immobilized lipase. *Enzym Microb Technol* 6: 443–446

6 Oladepo DK, Halling PJ, Larsen VF (1995) Effect of different supports on the reaction rate of *Rhizomucor miehei* lipase in organic media. *Biocatal Biotrans* 12: 47–54

7   Xu H, Li M, He B (1995) Immobilization of *Candida cylindracea* lipase on methyl acrylate-divinyl benzene copolymer and its derivatives. *Enzym Microb Technol* 17: 194–199

8   Bosley JA, Peilow AD (1997) Immobilization of lipases on porous polypropylene: Reduction in esterification efficiency at low loading. *JAOCS* 74 (2): 107–111

9   Gitlesen T, Bauer M, Adlercreutz P (1997) Adsorption of lipase on polypropylene powder. *Biochim Biophys Acta* 1345: 188–196

10 Jensen BF, Eigtved P (1990) Safety aspects of microbial enzyme technology, exemplified by the safety asessment of an immobilized lipase preparation, Lipozyme(tm). *Food Biotechnol* 4: 699–725

11 Kosugi Y, Takahashi K, Lopez C (1995) Large-scale immobilization of lipase from *Pseudomonas fluorescens* biotype I and an application for sardine oil hydrolysis. *JAOCS* 72 (11): 1281–1285

12 Gandhi NN, Vijayalakshmi V, Sawant SB, Joshi JB (1996) Immobilization of *Mucor miehei* lipase on ion-exchange resins. *Chem Eng J (Lausanne)* 61(2): 149–156

13 Stark M -B, Holmberg K (1989) Covalent immobilization of lipase in organic solvents. *Biotech Bioeng* 34: 942–950

14 Cho S -W, Rhee JS (1993) Immobilization of lipase for effective interesterification of fats and oils in organic solvent. *Biotech Bioeng* 41: 204–210

15 Sato S, Murakata T, Suzuki T et al. (1997) Esterification activity in organic medium of lipase immobilized on silicas with differently controlled pore size distribution. *J Chem Eng Jpn* 30(4): 654–661

16 Yokozeki K, Yamanaka S, Takinami K et al. (1982) Application of immobilized lipase to regio-specific interesterification of triglyceride in organic solvent. *Europ J Appl Microbiol Biotechnol* 14: 1–5

17 Reetz MT (1997) Entrapment of biocatalysts in hydrophobic sol-gel materials for use in organic chemistry. *Adv Mater* 9(12): 943–954

18 Lalonde J, Govardhan C, Khalaf N et al. (1995) Cross-linked crystals of *Candida rugosa* lipase: Highly efficient catalysts for the resolution of chiral esters. *J Am Chem Soc* 117(26): 6845–6852

19 Adlercreutz P, Barros R, Wehtje E (1996) Immobilization of enzymes for use in organic media. *Ann NY Acad Sci* 799: 197–200

20 Bosley JA, Clayton JC (1994) Blueprint for a lipase support: Use of hydrophobic controlled-pore glasses as model systems. *Biotech Bioeng* 43: 934–938

21 Analytical Method AF 95.1/3-GB. (1983) Novo Nordisk, Copenhagen, Denmark

22 Montero S, Blanco A, Virto MD et al. (1993) Immobilization of *Candida rugosa* lipase and some properties of the immobilized enzyme. *Enzyme Microbiol Technol* 15: 239–247

23 Huang FC, Ju YH (1994) Improved activity of by vacuum drying on hydrophobic microporous support. *Biotechnol Tech* 8: 827–830

24 Bailie PM, McNerlan SE, Robinson E, Murphy WR (1995) The immobilization of lipases for the hydrolysis of fats and oils. *IChemE* 73 : 71–76

25 Bosley JA, Peilow AD (1991) Supported enzyme. *European Patent Application EP* 424,130

26 Hansen TT, Eigtved P (1985) A new immobilized lipase for interesterification and ester synthesis. *Proc World Conf Emerging Technol in Fats and Oils Ind, Canne, France,* 365–69 Publ AOCS

27 Macrae AR (1985) Interesterification of fats and oils. In: J.Tramper, H.C. Van der Plas, P. Linko (eds): *Biocatalysts in organic synthesis.* Proceedings of an international Symposium held at Noordwijkerhout, The Netherlands, 14–17 April 1985. Elsevier, Amsterdam, 195–208

28 Tanaka T, Ono E, Takinami K (1981) *US patent 4,275,011*

29 Ajinomoto KK (1991) *Japanese patent application 03–183480*

30 Kao KK (1988) *Japanese patent application 63–214184*

31 Reetz MT, Zonta A, Simpelkamp J (1995) Efficient heterogenous biocatalysts by entrapment of lipases in hydrophobic sol-gel materials. *Angew Chem Int Ed Engl* 34: 301

32 Valivety R, Halling PJ, Macrae AR (1992) *Rhizomucor miehei* lipase remains highly active at water activity below 0.0001. *FEBS Lett* 301 (3): 258–260

33 Valivety RH, Halling PJ, Peilow AD, Macrae AR (1992) Lipases from different sources vary widely in dependence of catalytic activity on water activity *Biochim Biophys Acta* 1122: 143–146

34 Oladepo DK, Halling PJ, Larsen VF (1995) Effect of different supports on the reaction rate of *Rhizomucor miehei* lipase in organic media. *Biocatal Biotrans* 12: 47–54

35 Valivety RH, Halling PJ, Macrae AR (1992) Water effects on suspended lipase in organic solvents: Better characterised by thermodynamic activity rather than content. *Indian J Chem* 31B: 914–916

36 Gubicza L, Szak.cs-Schmidt A (1995) Production of bioflavours by lipase-catalysed esterification in organic solvents. *Med Fac Landbouww Univ Gent* 60 (4a): 1977–1982

37 Kvittingen L, Sjursnes B, Anthonsen T, Halling P (1992) Use of salt hydrates to buffer optimal water level during lipase catalysed synthesis in organic media: A practical procedure for organic chemists. *Tetrahedron* 48 (13): 2793–2802

38 Wehtje E, Svensson I, Adlercreutz P, Mattiasson B (1993) Continuous control of water activity during biocatalysis in organic media. *Biotechnol Techn* 7 (12): 873–878

39 Miller C, Austin H, Posorske L, Gonzlez J (1988) Characteristics of an immobilized lipase for the commercial synthesis of esters. *JAOCS* 65 (6): 927–931

40 Kwon SJ, Song KM, Hong WH, Rhee JS (1995) Removal of water produced from lipase-catalysed esterification in organic solvent by pervaporation. *Biotech Bioeng* 46: 393–395

5 Applications of Enzymes and Membrane Technology in Fat and Oil Processing

Mitsutoshi Nakajima, Jonathan B. Snape and Sunil Kumar Khare

Contents

Methods and Tools in Biosciences and Medicine
Methods in non-aqueous enzymology, ed. by M. N. Gupta
© 2000 Birkhäuser Verlag Basel/Switzerland

1   Introduction

The total world production of agricultural oils in 1990 was estimated at 80 million tons (MT) of which approximately 60 MT were vegetable oils, 18.6 MT animal oils and 1.4 MT fish oils [1]. This figure is rising and is expected to be over 105 MT by the year 2000, establishing agricultural oil processing as one of the most important sectors of food processing. Approximately 80% of all agricultural oils are used for food applications. The remaining 20% are used in industrial applications including detergents or soaps, cosmetics, lubricants, and carriers for sprays, paints, varnishes and plastics [2, 3]. However, with a few exceptions, the crude oil cannot be used. In order to obtain a product that has suitable properties for the applications, the crude oil needs to be refined. After the oil is refined it is sometimes modified in order to obtain a product with the desired physical and chemical properties as per intended use. Thus there is interest in the restructuring of fats and oils with respect to their fatty acid composition for nutritional and pharmaceutical applications. Conventionally, this has been done either by physical blending of fats and oils of desired type or by chemical catalysis. These processes require high temperatures and fatty acids are randomized making it difficult to obtain the product with required properties. For this reason enzyme-based processes that can be carried out under moderate reaction conditions and offer specificity in obtaining desired products with little or no side products are being developed at a rapid rate. Lipases are one of the most important enzymes in this regard and lipase-catalyzed hydrolysis, esterification and interesterification reactions have been extensively studied for applications in processing and modification of fats and oils [4, 5]. Membrane technology is gradually gaining acceptance in food processing [6] and also has potential applications in edible oil processing [7, 8]. Again, one of the main advantages of this technology over conventional processing is that much lower temperatures can be used. Since agricultural oils are susceptible to thermal damage, keeping the processing temperature to a minimum is advantageous besides saving the cost in energy.

1.1   Lipases

It is relevant here to describe some of the salient feature of the lipases to give an overall idea about their applications in the latter part of this chapter. However, detailed enzymatic and structural properties of lipase are covered in some recent reviews [9–11]. Lipases (triacyl glycerol acylhydrolase EC 3.1.13) belong to the hydrolase class of enzymes, attacking carboxyl ester bonds. These are widely distributed in animals (digestive tract and tissues), plants and microorganisms including bacteria, fungi and yeast. Porcine pancreas lipase is the best characterized lipase in terms of active site, structure, kinetic properties and reaction mechanism. It is found in two isozyme forms (lipase A and lipase B); both are glycoproteins with quite similar polypeptide chains (mol. wt. 11,000). Histidine is implicated to be involved in active site. The enzyme has a broad spectrum of side chain specificity. Calcium is required for activity whereas Zn^{++}, Cu^{++} and Hg^{++} inhibit the enzyme [12]. According to their specificity, microbial lipases have been classified into two broad categories, random type *(Candida rugosa, Corynebacterium acnes* and *Staphylococcus aureus)* and 1, 3 positional specific type *(Aspergillus* sp., *Mucor* sp. and *Rhizopus* sp.). *Geotrichum candidum* lipase shows specificity for the type of fatty acid, and cleaves fatty acid containing a cis bond at the 9th position of the chain [13]. Among various microbial enzymes *Mucor miehei, C. rugosa, C. antarctica* and *Rhizopus* sp. enzymes have been used in a large variety of industrial applications. Immobilized *M. miehei* with the trade name of Lipozyme is commercially available from Novo-Nordisk Co., Denmark.

The primary reaction of lipases is hydrolysis of triglycerides into corresponding fatty acids and glycerol. However, at low water content reaction equilibrium is shifted toward synthesis, thus esterification and transesterification reactions predominate. Esterification is the reversal of hydrolysis whereas in interesterification the acyl group is transferred from one ester to either an acid (acidolysis) or to an alcohol (alcoholysis) or other ester (transesterification).

1.2   Applications of lipases

Lipase based hydrolysis
Fatty acids and glycerol components of fats and oils are desired ingredients in soap and detergents, foods, pharmaceuticals and cosmetics. The average industrial production of fatty acids is reported to be 5.2 MT per annum, reflecting the industrial need of fat and oil hydrolysis [14]. Animal fats and vegetable oils such as palm, soybean and olive oils are commonly hydrolyzed for the above purposes. *Candida rugosa, Rhizopus delemer*, and *Mucor miehei* lipases have generally been used for the hydrolysis of fats and oils [15–18]. Lipase catalyzed hydrolysis of butter oil for production of characteristic flavor (due to free fatty acids) is one such major application [19–21]. Lipases are also used in fish oil hydrolysis for obtaining polyunsaturated fatty acids (PUFA) such as docosahexaenoic acid (DHA) and eicosapentaenoic acid (EPA).

Esterification
The importance of the esterification reaction lies in the fact that various types of such esters form the basic material for the flavor and fragrance industry. Lipase-based syntheses of a number of products have been widely worked out and reported in literature [6, 7] viz. ethyl butyrate for apricot and peach flavor [22], isoamyl butyrate for pine [23], benzoyl laureate for musk aroma [24], terpene alcohol with various low chain acid especially geranyl and citronellyl occurring in essential oils [25]. Mainly *Candida rugosa, Rhizopus* sp., *Mucor miehei*, and *Pseudomonas* sp. lipases have been used for the above applications.

Interesterification

Interesterification reactions of lipases offer enormous scope for industrial applications. Some industrial processes where lipase interesterification reactions are being used are

- Production of cocoa butter equivalent 1,3 saturated (palmitic/stearic) 2- oleloyl glyceride by tranesterification of palm oil or sunflower oil [26];

- Human milk fat substitute Betapol by lipase-catalyzed transesterification between tripalmitin and oleic acid [27];

- Pharmaceutically important structured lipids such as poly-unsaturated fatty acid rich lipids by incorporating PUFA (especially DHA and EPA from fish oil) into vegetable oils, i.e., soybean, groundnut, canola and peanut by using *Mucor miehei* lipase [28, 29].

- Low calorie structured lipids such as medium chain triglycerides are designer fats obtained by replacing normally occurring long chain fatty acids by low/medium chain fatty acid e.g. caprylic (C8:0) and capric acid (C10:0). The idea is to reduce the calorie intake and also to use this as therapeutic food for people suffering with the problem of absorption of long chain fatty acid such as in the case in cystic fibrosis [30]

- Biodiesel from vegetable oils: fatty acid esters of vegetable oils having reduced viscosity prepared conventionally by pyrolysis, microemulsifiacation and alkali-catalyzed reaction are reported to be useful as biodiesel [31]. Similar fatty acid esters are produced from soybean or canola oils enzymatically (*Mucor miehei* or *Candida antarctica* lipase).

1.3 Surfactant modified lipase

This chapter describes the surfactant modification of enzymes as a simple and easy method to obtain a lipase preparation that solubilizes well, evenly disperses in organic solvent and gives higher interesterifiaction activity with 1,3-positional specificity. Lipase is modified by complexing with surfactant sorbitan monostearate or stearic acids so that the hydrophobic moieties of surfactants arrange themselves on the outer side of the enzyme.

1.4 Application of membrane technology to lipase catalyzed reactions

The cost of lipases is relatively high so it is desirable for the enzyme to be reused. Apart from immobilized enzymes, membrane bioreactors have also been used for the modification of fats and oils.

The kinetics and mechanisms of immobilized lipases [32], the use of immobilized lipase reactors [33], immobilization onto cellulosic materials [34] and biochemical engineering aspect of the enzymatic modifications of fats and oils, including some useful informations on membrane reactors [35, 36] have been extensively reviewed in recent years.

Membrane processes can be divided into three types:

- Two-phase membrane reactor systems: In two-phase membrane reactors the lipase is immobilized either onto the surface of the membrane or within the membrane. For the reactions, the organic phase is on one side of the membrane and the aqueous phase on the other. Diffusion of the two phases through the membrane to the lipase is necessary for hydrolysis to occur. The advantage of this system is that emulsion formation is not needed and at the sametime the products can be separated. The conditions on either side of the membrane can be optimized and the feed streams can be recycled, so that high conversion efficiencies can be achieved. The main disadvantage is that the reaction rate is limited by the membrane surface area and the mass transfer rate across the membrane, which necessitates the use of large membrane surface areas, leading to increased costs for this system.

- Emulsion-phase membrane reactor systems: The advantage of this type of reactor is that the reaction rate is not limited by the membrane area but by the area of interface between oil and water phases. Large interfacial areas can be achieved by high intensity mixing. One disadvantage is that the emulsion is required to be broken after completion of the reaction.

- Single-phase reactor systems: The operating costs are comparatively less. In addition, the reactor systems are very simple in design, thus capital cost is reduced. The main disadvantage, however, is that there is no separation of the products from the reactants. The surfactant modified lipases and their operation in hollow fibre membrane reactor described in the methods section fall into the single-phase reactor category.

2   Materials

Chemicals

Cetyl alcohol	Sigma
1,2- dipalmitoyl	Unilever
1,3-distearoyl-2-palmitoyl glycerol	Unilever
Hexadecane	Sigma
Mono-, di-, and tripalmitin	Sigma
Mono-, di-, and tristearin	Sigma
Palmitic acid	Sigma
Sorbitan mono-stearate	NOF Corporation

Stearic acid	Sigma
3-stearoyl glycerol	Unilever
Tripalmitin	Sigma

Enzymes
- Lipase MF-30: *Pseudomonas* sp. from Amano Pharmaceuticals Co., Japan.
- Lipase Saiken 100: *Rhizopus japonicus* from Nagase Chemical Co., Japan
- Lipase Asahi: *Chromobacterium viscosum* from Asahi Co. Japan.

Solutions
- Acetate, 0.05M, pH 5.6
- Citrate phosphate, 0.1 M, pH 7.0
- Dry n-Hexane: HPLC grade hexane is used after drying over 5 Å molecular sieve 250 g/5 L to give a water concentration of 10 mg/L.

Equipment
- Membrane filter: LCR 13, LH, Milipore Co. USA
- Polyamide flat sheet membrane reactor system: NTU-4208, molecular weight cut off (MWCO)-8000 (Nitto Denko Corp., Japan). The membrane was used in a stirred cell (UHP-62 K, filter size 62 mm, total membrane area 27 cm^2, total volume capacity 200 cm^3, Advantec Toyo, Japan)
- Hollow fiber membrane reactor system: Chiyoda Seisakusho Co. Ltd., Japan. Resistant to hydrophobic organic solvent and the membrane module (Toray Co., Japan). The module is made of polyphenylene sulphone (PPSO) membrane having MWCO of 50 kD. This membrane consisted of bundle of 63 fibers, each 13.5 cm in length, 1.2 mm outer diameter and 0.8 mm inner diameter. The membrane surface area was 32 cm^2.
- Automatic protein analyzer: Model FP-428, LECO, USA
- Lyophilizer: Vertis Lab model
- Centrifuge: Beckman Hi-speed model
- Automatic pH stat/titrator: Metrohm, Switzerland and Karl Fisher Titrator 684 KF
- Coulometer: Metrohm, Switzerland
- Gas Chromatograph: Shimadzu Corp. Model GC-6A with Flame Ionization Detector (FID), capillary column 0.25 mm internal diameter, 25 m column length, 0.1 µm film thickness. Silicon OV-1 Glass column (3 mm x 0.5 m) (GL Science, Tokyo, Japan).

3   Methods

3.1   Protocols

Protocol 1   Surfactant modification of lipases [37]

1. Crude *Pseudomonas* sp. lipase MF-30 or lipase Saiken 100 (*Rhizopus japonicus*) (500 mg) is dissolved in 500 mL acetic buffer (0.05 N, pH 5.6).
2. Surfactant solution of sorbitan mono-stearate (50 mg/mL) is prepared either in ethanol or water.
3. Surfactant is added into the magnetically-stirred enzyme solution.
4. The mixture is kept at 50 °C with constant stirring for 1 day.
5. The resultant precipitate is collected by centrifugation (5000 g, 4°C, 15 min).
6. Lyophilize the precipitate to obtain modified lipase powder.

Protocol 2   Lipase modification with stearic acid by two-phase method [38]

1. Lipase Saiken (*Rhizopus japonicus*) or Asahi (*Chromobacterium viscosum*) (500 mg) is dissolved in 40 mL of citrate-phosphate buffer (0.1 M; pH 7.0).
2. Stearic acid is dissolved in 40 mL of *n*-hexane.
3. The lipase and stearic acid solution are mixed and incubated at 4 °C for 24 hrs with constant stirring.
4. The mixture is centrifuged at 3000 g at 4 °C for 10 min.
5. The upper hexane phase is carefully removed by Pasteur pipette after centrifugation.
6. Residual hexane is evaporated by blowing the nitrogen at surface for 1hr.
7. The modified lipase is obtained as precipitate gel floating at the top of aqueous buffer phase.
8. This is recovered by centrifugation at 8,000 g at 4 °C for 15 min.
9. The modified lipase is lyophilized and stored in a desiccator at 4 °C.

Protocol 3   Palm oil hydrolysis by surfactant modified lipase [37]

1. Mix 50 mg of surfactant-modified lipase, 25 ml acetate buffer (0.05M, pH 5.6), 5 mL palm oil and 5 mg calcium chloride.

2. The mixture is constantly stirred in a magnetic stirrer at 500 rpm.

3. Monitor the release of free fatty acid by continuous titration with NaOH (0.1 M) using a pH-stat.

Protocol 4   Wax ester synthesis in a membrane reactor with modified lipase [39]

1. Polyamide flat sheet membrane (NTU-4208, molecular weight cut off (MWCO)-8000 is fitted in a stirred cell (UHP-62 K, filter size 62 mm, total membrane area 27 cm^2, total volume capacity 200 cm^3). A schematic illustration of the system is shown in Figure 1.

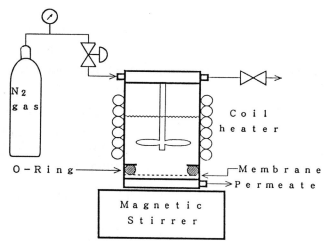

Figure 1   A schematic illustration of the membrane reactor system for the repeated batch esterification. Membrane: NTU-4028 (polyimide, molecular weight cut off (MWCO) 8000; Nitto Denko Corporation, Osaka, japan). Reactor: UHP-62K (membrane area 27 cm^2, total volume capacity 200 cm^3, Advantec Toyo, Tokyo, Japan).

2. The membrane is washed sequentially with water, ethanol and hexane before each use.

3. Esterification of cetyl alcohol with palmitic acid in the batch reaction is carried out in membrane reactor with 25 mmol/L cetyl alcohol, 25 mmol/L palmitic acid, 40 mg of surfactant modified lipase (Saiken 100) and 25 mL hexane.

4. The reaction mixture is stirred at 500 rpm and temperature maintained at 50 °C by means of a coil heater.

5. After 30 min of reaction the solution is filtered by applying pressure with nitrogen gas, and the permeate flux is measured.

6. The composition of permeate is analyzed by gas chromatography for fatty acid and wax ester concentration.

Repeated batch reaction

1. The repeated batch reaction is performed by adding fresh substrate solution (25 mmol/L cetyl alcohol, 25 mmol/L palmitic acid in *n*-hexane) to the reactor after filtration of reaction mixture in the first run.

2. The operation is repeated similarly. The permeate flux and permeate composition is monitored after each run.

3. The membrane surface is washed with sponge and water after last run.

Analysis of the ester product

1. The concentration of ester formation is monitored in the permeate by gas chromatography using silicon OV-1 (3mm x 0.5 m glass column).

2. The analysis is performed under following conditions: injection temperature, 250 °C; oven temperature initial 110 °C, 7 °C/min up to 250 °C; carrier N_2; flow rate 5.0 mL/min.

Protocol 5 Interesterification of triglycerides and fatty acids by lipase in a hollow-fiber reactor [40]

1. n-Hexane is treated with molecular sieve to decrease the water concentration to 10 mg/L or below.

2. For interesterifiaction in solvent system, tripalmitin (PPP) and stearic acid (S) are dissolved in 50 mL dry n-hexane.

3. In this reaction mixture 30 mg of modified lipase (Saiken 100) is added and the reaction is carried at 40 °C for 3 hrs with magnetic stirring at 500 rpm.

4. For solvent free system, substrates 5 g of PPP and 5 g of S are melted at 75 °C, followed by the addition of 300 mg enzyme.

5. Reaction is carried out at 75 °C for 3 hrs with constant stirring at 500 rpm on a magnetic stirrer.

6. The interesterification reaction products, 1,2-dipalmitoyl-3-stearoyl glycerol (PPS) and 1,3-distearoyl-2-palmitoyl glycerol (SPS) are monitored by gas chromatography.

The hollow fiber membrane reactor system

1. The flow diagram of reactor system is shown in Figure 2. This system can be operated batch wise or continuously.

Batchwise operations

2. A solution of 660 ml n-hexane containing 6 mmol/L (PPP) and 12 mmol/L (S) kept at constant magnetic stirring at 800 rpm is recirculated in the system for 1 hr through the hollow fiber membrane module.

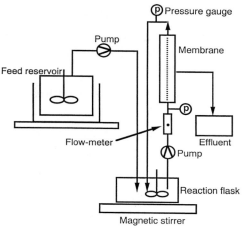

Figure 2 Flow diagram of hollow membrane fibre reactor system

3. The flow rate at 200 mL/min and temperature at 40°C are maintained throughout.

4. The reaction is initiated by adding 240 mg modified lipase (Saiken 100) into the recirculating solution.

5. After 20 hrs of recirculation the outlet of membrane module is opened and 0.5 bar transmembrane pressure is applied in order to force 400 ml of solution to permeate through membrane (average flux was 10.4 mL/ (m^2s). The membrane rejects the biocatalyst while the substrates and the products remain dissolved in hexane permeate.

6. The system is purged with 400 ml *n*-hexane and after that 400 ml of recirculating solution is forced to permeate through the membrane module.

7. The process is repeated once more and then, 400 ml n-hexane solution containing 6 mmol/L PPP and 12 mmol/L S is again added to the remaining solution in the system.

8. The process is repeated three times using lipase retained in the system.

Continuous operation

9. In case of continuous operation, the system is also operated similarly. The feed rate of substrate solution is kept at 0.75 mL/min (residence time 14.7 hrs) for 240 mg of enzyme used.

10. The formation of transesterification reaction products PPS and SPS is monitored by gas chromatography.

Estimation of interestification products by gas chromatography

1. The interesterification reaction products, 1,2-dipalmitoyl-3-stearoyl glycerol (PPS) and 1,3-distearoyl-2-palmitoyl glycerol (SPS) are monitored by gas chromatography.

2. The samples withdrawn from reaction mixture are filtered with 0.5 µm Millipore filter.

3. The filtered samples are dried under nitrogen at 60 °C.

4. The dried samples dissolved in 1 mL of hexane containing internal standard (1 mg hexadecane/mL) are analyzed by gas chromatography equipped with capillary column under following conditions.

5. Operating temperature 80–360 °C is programmed at the rate 10 °C/min; injector and detector (FID) at 370 and 400 °C.

4   Troubleshooting

- *Temperature:* The yield of modified lipase sorbitan monostearate complex (Protocol 1) is temperature dependent. Lower yield is observed at temperatures below 50 ° C, while raising temperature beyond 55°C leads to loss of enzyme activity due to heat inactivation.

- *Surfactants*: Anionic surfactants do not form lipase surfactant complex (LSC), cationic surfactants form LSC but with relatively low activity, nonionic surfactants are found to be most suitable for LSC formation.

- *Stearic acid solution*: While using stearic acid in hexane for modification (Protocol 2), it is difficult to dissolve it in n-hexane at room temperature, however raising the temperature to about 40 °C dissolves it, giving a clear solution.

- *Fragility of modified lipase*: The modified lipase obtained at the interface of hexane (having stearic acid) and buffer (having lipase) is sometimes very fragile (Protocol 2). It should be carefully removed after centrifugation using a spatula. Vigorous handling breaks it into small pieces that are very light in weight and float on surface making it difficult to collect and hence, reducing the yield.

- *Membranes*: In the application of lipases using a membrane reactor, the selection of membrane is important (Protocols 4 and 5), generally, hydrophobic membranes such as PTFE are found to be more suitable.

- *Decrease in permeate flux:* During the wax ester synthesis using LSC in membrane reactor system (Protocol 4), decrease in permeate flux is observed from first run (34 L/(m^2 · h · kg/m^2)) to 10 L/(m^2 · h · kg/m^2)) in the fifth run. This is mainly due to fouling. The sponge washing may be attempted which restores the flux recovery to 70–80%.

- *Concentration polarization:* It is suggested that oil concentration and permeate fluxes should be kept relatively low, to avoid the problem of concentration polarization in oil processing by using lipases in membrane reactors.

5   Applications

5.1   Application of modified lipase reactions

Hydrolysis activity of surfactant modified lipase
The hydrolysis of palm oil by surfactant modified lipase is described in Protocol 3. The effect of sorbitan mono stearate modification on hydrolysis activitiy of *Psuedomonas* sp. lipase, is shown in Table 1. *Pseudomonas* sp. lipase-sorbitan mono stearate complex hydrolyzed the palm oil most efficiently.

Table 1   Effect of surfactant modification on hydrolysis activity of *Pseudomonas* sp. lipase

Surfactants	Hydrophilic Group	Hydrophobic Group	HLB	Yield of modified lipase LSC (%)	Hydrolysis activity μmol/(min · g LSC)
DKester F-20	Sucrose	Stearic acid	2.0	10.7	43.9
Lauric acid SE	Sucrose	Lauric acid	4.0	11.9	6.0
Nonion SP-60R	Sorbitan	Stearic acid	4.7	28.8	11.9

LSC: Lipase Surfactant Complex

Esterification activity of surfactant modified lipase
Esterification activity of sorbitan monostearate modified *R. japonicus* lipase and its application in wax ester synthesis are described in Protocol 4. Modification of lipase by surfactant sorbitan monostearate led to significant increase in esterification activities. The observed activity 37.1 mmol/(min · mg protein) [39] was much higher as compared to other modified lipase preparations such as 25 in case of ion exchange bound lipase [46] and 28.4 in case of PEG modified lipase [47]. This is probably due to better dispersibility of LSC having more hydrophobic surface and more contact probability between LSC and hydrophobic substrates cetyl alcohol and palmitate.

Wax esters are industrially important products used as petroleum lubricants, antifoaming agents and ingredients in cosmetics. Currently these are manufactured by reacting alcohol with fatty acid at high temperature (up to 250 °C), in presence of tin, titanium or sulfuric acid catalyst up to 20 hrs. Cetyl palmitate (Protocol 4) is a principal ingredient of whale oil [48] and in much demand due to short supply of natural whale oil. As high as 90% conversion was achieved in 30 min and this increased to 96% in 1hr, reaction time (Fig. 3). The same rate of conversion is maintained up to ten repeated batch operations.

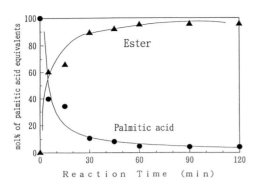

Figure 3 Time course of the esterification of palmitic acid and cetyl alcohol. The reaction was carried out with 25mmol/L cetyl alcohol, 25 mmol/L palmitic acid and 40 mg of modified lipase at 500 rpm, 50 °C.

The ultrafiltration system used in the present study was suitable for retaining the LSC in reaction system and operation of lipase-catalyzed esterification. Lipase-based synthesis of number of such products have been reported in literature [4,5].

Interesterification activity of surfactant modified lipase
Interesterification by lipase is favored when the concentration of water in the reaction is very low. This reaction is currently of great industrial interest to change the physiochemical properties of oils and fats for high economical value. Lipases often show operation constraints and result into side reaction products such as diglycerides and mono glycerides, therefore reaction specificity for 1, 3 position remains desirable. Surfactant and stearic acid modification of lipase described in Protocols 1 and 2 impart these desirable attributes to the lipase. The modified lipase was tested for intereserification reaction between the stearic and tripalmitin in *n*-hexane following the procedures as described in Protocol 5. The PPS, SPS and SSS were the likely products besides some MG and DG. A typical reaction course of interesterification by modified lipase under above condition is shown in Figure 4. It can be clearly seen that concentration of substrates tripalmitin and stearic acid decreased with time whereas the

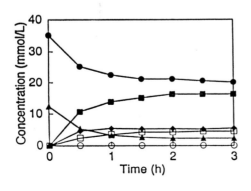

Figure 4 Interesterification reaction time course for the modified lipase obtained from Lipase Saiken 100 (*Rhizopus japonicus*; Nagase Biochemicals, Kyoto, Japan) and Emazol S-10(F) (sorbitan monostearate; Kao Corp., Tokyo, Japan). Substrates were 0.5 g of tripalmitin and 0.5 g of stearic acid. The amount of modified lipase was 30 mg. The reaction was carried out at 500 rpm stirring and 40 °C in n-hexane system. Palmitic acid (■), stearic acid (●), tripalmitin (▲), 1,2-dipalmitoyl-3-stearoyl glycerol (PPS) (♦), 1,3-distearol-2-palmitoyl glycerol (SPS) (□); tristearin (SSS) (○).

products PPS and SPS increased with time as estimated by GC. The efficiency of conversion was in the range of 80–90%.

The SSS was not produced at all which shows that the modified lipase has 1, 3- specificity. MG were also not produced and DG were observed to be less than 6% of TG. The reaction reached steady state after 2 hrs. The preparation should be useful in some of the other interesterification applications as well.

However, the operation of interesterification reaction in industrial processes neccessiates their use in appropriate reactor system. Use of membrane bioreactors is an attractive choice in this regard.

5.2 Application of membrane technology

The wax ester synthesis by esterification reaction between oleic acid and cetyl alcohol in n-hexane and interesterification of tripalmitin with stearic acid using modified lipase in hollow fiber reactor, a single phase type reactor system, was studied as model system to assess the suitability of modified lipase in membrane reactor operations.

The choice of membrane in Protocol 4 was made because of its stability in hexane. The conversion was in the range of 80–90% during 10 operations of a repeated-batch experiment. Neither deactivation of enzyme in membrane reactor system nor any permeation of LSC was observed.

The next Protocol 5 describes the use of hollow fiber membrane reactor in batch as well as continuous mode for interesterification reaction triglycerides and fatty acids catalyzed by stearic acid modified *R. japonicum* lipase in n-hexane. The PPSO membrane recently developed by Toray Co. Ltd. (Tokyo, Japan) was used. This has been made by the oxidation of polyphenylene sulphide. It was chosen because of its high stability in organic solvents such as hexane, ethyl acetate and pyridine. The module consisted of a bundle of 63 fibers of 13.5 cm length, 1.2 mm outer diameter and 0.8 mm inner diameter. The membrane surface area of the module was 32 cm^2. This system could be operated either batch-wise or continuously by opening or closing the outlet of the membrane module (Fig. 2). The PPS and SPS was the only reaction products obtained because of the 1,3 specificity of lipase. Figure 5 shows that the modified lipase used in four runs catalyzed predominantly the interesterification reaction to produce PPS and SPS. The steady state in the first run reached after 2 hrs and thereafter no substantial change in the product concentrations was observed. The steady state in rest of the three runs was reached after 10 hrs. The modified lipase in continuous operation retained its original activity for more than 72 hrs of operation (Fig. 6). The membrane was stable and showed no leakage or swelling. The enzyme was retained by the membrane but hexane and oil could freely permeate. The reactor was operated in a continuous mode with a residence time of almost 15 hrs, long enough to ensure the system was

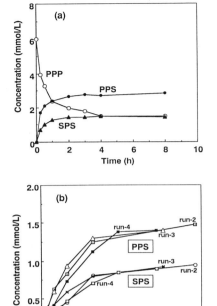

Figure 5 (a) Concentration profiles of triglycerides in run 1 using the hollow-fiber membrane reactor system. The reaction was carried out in 660 mL n-hexane containing 6mmol/L tripalmitin and 12 mmol/L stearic acid. (b). Concentration profiles of interesterification products, PPS and SPS, in runs 2,3 & 4. The reaction solution was magnetically stirred at 800 rpm and recirculated through the membrane module at 200 mL/min. The temperature of the system was maintained at 40 °C.

at steady state. About 10% DG was produced in batch and continuous membrane reactor as compared to 6% in stirred tank reactor.

Overall, membrane system offers a potential alternative to stirred tank and fixed bed reactor or nonaqueous enzyme operations. Prazeres et al. (1993) used a ceramic UF membrane (Carbosep, Rhone-Poulenc) with a nominal MWCO of 10 kDa to hydrolyze olive oil [49]. A subsequent study with the same reactor demonstrated the feasibility of a continuous process [50]. Despite these successful reports, however, Prazeres et al. (1994b) expressed doubts about the successful use of UF membranes in conjunction with reversed micelles due to the permeation of surfactant monomers causing contamination in the product stream [51]. It also affected the dynamic equilibrium between the reversed micelles and the monomers. An additional concern was the stability of reversed micelles in the high shear environment often encountered in membrane reactors. It remains to be seen if these problems can be overcome and if membrane technology can form a viable process for lipase applications in oils and fats processing industries.

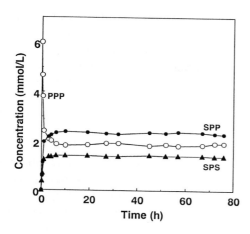

Figure 6 Operational stability of the modified lipase in the continuous hollow-fiber membrane reactor system. The reaction flask was filled with 660 mL stearic acid. The feed rate was 0.75 mL/min of substrate solution. The reaction solution was operated under the same conditions as in Fig. 5.

6   Remarks and Conclusions

6.1   Modification of lipases

The modification of lipase to obtain a reusable preparation that can be solubilized in an organic solvent and which gives acceptable reaction rate is one of the important areas in the research on lipases. Several methods have been tried for the modification of enzyme surface to impart additional hydrophobicity making them soluble in solvents.

6.2   Covalent modifications

Polyethylene glycol modification is a well known approach for modifying enzymes for their use in organic solvents [41]. Polystyrene and polyacrylates [42], bis-imido esters [43], other lipids and lipid derivatives have also been used for covalent modifications.

6.3   Non-covalent modifications

Enzyme surfactant complexes or enzyme polymer complexes described in Protocol 1 of this chapter fall in the category of non-covalent type modifications. Non-ionic or cationic surfactants form such solvent soluble complexes, however anionic surfactants do not yield such products.

Results obtained with surfactant modification of lipase (Protocol 1) and its effect on hydrolysis activity are shown in Table 1.

The modification of enzyme by surfactants depends on the chemical properties of surfactants determined by the type of hydrophobic and hydrophilic groups, and hydrophobic-lipophilic balance (HLB). The enzyme surfactant ratio also affects the LSC yield and activity during lipase modification [37].

6.4   Modification of lipases by stearic acid

The modification can be carried out either by drop-wise method or two phase method. In drop-wise method the lipid is dissolved in a polar solvent and added drop-wise into enzyme solution)in buffer. A typical procedure for two phase modification of a lipase by stearic acid in two phase method is described in Protocol 2. The crude lipase is dissolved in buffer and stearic acid dissolved in n-hexane is used as modifier; when mixed both remain separated as two different phases. After 24 hrs of incubation, the active lipase modified by stearic acid separated from the crude forming an insoluble layer at the interface. The modified lipase also acquired good amount of interesterification activity as compared to crude/native lipase which has very little or no such activity. Lower ionic strength and pH near neutrality favor the modification by stearic acid giving high yield and activity. Under optimum conditions of the modification interesterification activity (9.14 IU in n-hexane) increased 9–10 times over that of native which had very little or no interesterification activity (Tab. 2) [38]. Covalent modification of lipase with polyethylene glycol (PEG) has been reported to give 1.3 times increase in interesterification [44]. Lipid coating of lipase by diakyl amphiphiles didodecyl N-D-glucono L- glutamate gave 33% yield of modified lipase and 12.2 unit interesterification in benzene [45]. Modification by hydrophobic bisimido esters of varying chain length resulted in 3–6 time increase in interesterifiaction in benzene media [43].

Table 2   Characterization of native and stearic acid modified *C. viscosum* lipase

Prop.	Native	Modified
Total wt. (mg)	250	66
Protein (%)	2.39	6.17
TE activity (U)*	0	9.14
Hydrolysis activity (U)	6.8	4.65
Total activity recovered	0	60.05

Transesterification activity is defined as mmol PPS formed/(g protein · hr)
Hydrolysis activity is defined as µg p-nitrophenol released/(mg protein · min)

It has been demonstrated that enzyme and membrane technology has much potential in lipase applications and oil processing. However, much work remains to be done before widespread acceptance and implementation is achieved. In the case of membrane reactors it appears that hydrophobic membranes such as PTFE are more suitable to immobilize lipase.

In the majority of studies to date, the oil concentration has been kept low and the permeate fluxes have been relatively low, so problems of concentration polarization have not been reported. However, in actual industrial processes it is likely that concentration polarization will be an important factor.

There have been few long-term studies of the use of membranes for refining or modifying agricultural oils and therefore the problem of fouling has not been widely reported.

In fact, the design and operation of industrial scale membrane modules for use in agricultural oil processing has received inadequate attention. It is hoped that the experience gained in related industries can be applied successfully in this sector.

Advances in recombinant technology and bioseparation have led to the availability of lipases with desired specificity and stability. This in turn has led to the production of novel lipids from agricultural oils and further advances in membrane technology are required to enable these compounds to be processed under necessary mild conditions. In conclusion, there is vast potential for membrane technology in the agricultural oil industries but more fundamental research is required before this potential can be realized.

Acknowledgments

Part of the work described in the chapter was carried out by SKK while working as United Nations University (UNU) Fellow at National Food Research Institute (NFRI). The authors gratefully acknowledge the support provided by UNU, Tokyo; NFRI, Tsukuba and Central Institute of Agricultural Engineering, ICAR, New Delhi.

Abbreviations

DG   Diglycerides
DHA   Docosahexaenoic acid
EPA   Eicosapentaenoic acid
GC   Gas chromatography
HLB   Hydrophobic-lipophilic balance
HPLC   High performance liquid chromatography
LSC   Lipase surfactant complex
MG   Monoglycerides
MWCO   Molecular weight cut off
PEG   Polyethylene glycol
PPP   Tripalmitin
PPS   1,2-dipalmitoyl-3-stearoyl glycerol

PPSO   Polyphenylene sulphone
PTFE   Polytetrafluoroethane
PUFA   Polyunsaturated fatty acid
SPS   1,3-Distearoyl-2-palmitoyl glycerol
TG   Triglycerides
UF   Ultrafiltration

References

1 Bockisch M (1998) Fats and Oils Handbook, AOCS Press, Champaign, Illinois, 1-52
2 Kaufman J, Ruebusch RJ (1990) Oleochemicals – a look at world trends. *INFORM* 1: 1034–1048
3 Pryde EH, Rothfus JA (1989) Industrial and non food use of vegetable oils. In: G. Robbelen, RK Downey, A Ashri (eds): Oil Crops of the World: Their Breeding and Utilization. McGraw-Hill, New York
4 Vulfson EN (1994) Industrial applications of lipases. In: P Wooley, SB Peterson (eds) Lipases: Structure, biochemistry and applications. Cambridge University Press, London, 271–288
5 Bosley J (1997) Turning lipases into industrial biocatalyst. *Biochem Soc Transac* 25:174–178
6 Cuperus FP, Nijhuis HH (1993) Applications of membrane technology to food processing. *Trends Food Sci. Technol* 4:277–282
7 Koseoglu SS, Engelgau DE (1990) Membrane applications and research in the edible oil industry: An assessment. *J Am Oil Chem Soc* 67: 4, 239–249
8 Cheryan M, Raman LP, Rajagopalan N (1994) Refining vegetable oils by membrane technology. In: Yano, R Matsuno, K Nakamujra (eds): Developments in Food Engineering. Blakie, London, 677–679
9 Desnuelle P (1972) The lipases. In: Boyer PD (ed): The Enzymes, Vol.VII, 3rd ed., Academic Press, New York, 575–616
10 Wooley P, Peterson SB (eds) (1994) Lipases: Structure, biochemistry and applications. Cambridge University Press, London
11 Cambillau, C.(1996) Acyl glycerol hydrolases: Inhibitors, interface and catalysis. *Current Opinion in Structural Biol* 6: 449–455
12 Worthington V (1993) Worthington enzyme manual. Worthington Biochemical Corporation, New Jersey.
13 Macrae AR (1983) Lipase catalyzed interesterification of oils and fats. *J Am Oil Chem Soc* 60: 291–294
14 Hui YH (ed) (1996) Bailey's Industrial oils and fats products, vol.5. John Wiley, New York
15 Park YK, Pastore GM, DeAlmeida MM (1988) Hydrolysis of soybean oil by a combined lipase system. *J Am Oil Chem Soc* 65: 252–254
16 Yang D, Rhee JS (1992) Hydrolysis of olive oil by immobilized lipase in organic solvent. *J Am Oil Chem Soc* 40: 748–752
17 Holmberg K, Osterberg E (1988) Enzymatic preparation of mono glycerides in micro emulsions. *J Am Oil Chem Soc* 65: 1544–1548
18 Slaughter JC, Weatherley LR Wilkinson A (1993) Electrically enhanced enzymatic hydrolysis of vegetable oils using lipase from *Candida rugosa*. *Enzyme Microb Technol* 15: 293–296
19 Arnold RG, Shahani KM, Dwivedi BK (1975) Application of lipolytic enzymes to flavor development in dairy products. *J Dairy Sci* 58: 1127–1143
20 Kosikowski FV (1976) Flavor development by enzyme preparation in natural and processed cheddar cheese. United States Patent 3,975,544
21 Chen JP, Chang K-C (1993) Lipase catalyzed hydrolysis of milk fat in lecithin reverse micelles. *J Ferment Bioeng* 76: 98–104
22 Arctander S (1969) Application of esters in perfume chemistry. Weigner, Montclair.
23 Langrand G, Triantaphylides C, Baratti J (1988) Lipase catalyzed formation of flavor estrers. *Biotechnol Lett* 10: 549–554

24 Okazaki S-Y, Kamiya N, Abe K et al. (1997) Novel preparation method for surfactant lipase complexes utilizing water in oil emulsions. *Biotechnol Bioeng* 55: 455–460

25 Yee LN, Akoh CC, Phillips RS (1997) Lipase catalyzed transesterification of citronellyl butyrate and geranyl caproate: effect of reaction parameters *J Am Oil Chem Soc* 74: 255–260

26 Quinlan P, Moore S (1993) Modification of triglycerides by lipases: process technology and its application to the production of nutritionally improved fats. *INFORM* 4: 580–585

27 King DM, Padley FB (1990) Milk fat substitutes. European Patent No. 0 209 327

28 Newton I (1997) Meeting probes n-3 fatty acid medical role. *INFORM* 8: 176–180

29 Huang, K-H, Akoh, CC (1994) Lipase catalyzed incorporation of n-3 polyunsaturated fatty acids into vegetable oils. *J Am Oil Chem Soc* 71: 1277–1280

30 Timmerman F (1993) The medium chain triglycerides: the unconventional oil. *Int Food Ingred* 3: 11–18

31 Ali Y, Hanna MA (1994) Alternate diesel fuel from vegetable oils. *Bioresource Technol* 50: 153–163

32 Malcata FX, Reyes HR, Garcia HS et al. (1992) Kinetics and mechanisms of reactions catalyzed by immobilized lipases. *Enzyme Microbial Technol* 14: 426–446

33 Malcata, FX, Reyes, HR, Garcia et al. (1990) Immobilized lipase reactor for modification of fats and oils. *J Am Oil Chem Soc* 67: 890–910

34 Gemeiner P, Stefuca V, Bales V (1993) Biochemical engineering of biocatalysts immobilized on cellulosic materials. *Enzyme Microbial Technol* 15: 551–566

35 Buhler M, Wandrey C (1992) Oleochemicals by biochemical reactions? *Fat Sci. Technol* 94: 82–94

36 Snape JB, Nakjima M (1996) Processing of agricultural fats and oils using membrane technology. *J Food Eng* 30: 1–41

37 Isono Y, Nabetani H, Nakajima M (1996) Preparation of lipase surfactant complex for the catalysis triglyceride hydrolysis in heterogeneous reaction system. *Bioprocess Eng* 15: 133–137

38 Khare SK, Maruyama, T, Kuo TM, Nakajima, M (1998) Modification of *Chromobacterium viscosum* lipase by stearic acid in two phase method.

39 Isono Y, Nabetani H, Nakajima M (1995) Synthesis of wax ester using membrane reactor with lipase-surfactant complex in hexane. *J Am Oil Chem Soc* 72: 887–890

40 Basheer S, Mogi K, Nakajima M (1995) Development of a novel hollow-fiber membrane reaction for the interesterification of triglycerides and fatty acids using modified lipase. *Process Biochem* 30: 531–536

41 Inada Y, Matsushima A, Takahashi K, Saito Y (1990) Polyethylene glycol modified lipase soluble and active in orgqanic solvents. *Biocatalysis* 3: 317–328

42 Norin M, Boutelje J, Holmberg E, Hult K (1988) Lipase immobilized by adsorption. *Appl. Microbiol Biotechnol* 28: 527–530

43 Basri M, Ampon K, Yunus WMZ et al. (1992) Amidation of lipase with hydrophobic imidoesters. *J Am Oil Chem Soc* 69: 579–583

44 Kodera Y, Nishiumura H, Matsushima A et al. (1994) Immobilized lipase reactors for modifications of fats and oils. *J Am Oil Chem Soc* 71: 335–338

45 Goto M, Kameyama H, Miyata M, Nakashio F (1993) Design of surfactant suitable for surfactant coated enzyme as catalysts in organic media. *J Chem Eng Japan* 26: 109–111

46 Okahata Y, Ijiro K (1988) A lipid coated lipase as new catalyst for triglyceride synthesis in organic solvents. *J Chem Soc Commun* 1392–1394

47 Nishio T, Takahashi K, Tsuzuki T et al. (1988) PEG modified lipase for ester synthesis. *J Biotechnol* 8: 39–44

48 Abe Y (1988) *Yushi-Yuryou Handbook*, 1st ed., Saiwai Syobo, Tokyo

49 Prazeres DMF, Garcia FAP, Cabral JMS (1993) An ultrafiltration membrane bioreactor for the lipolysis of olive oil in reversed micellar media. *Biotechnol. Bioeng* 41: 761–770

50 Prazeres DMF, Garcia FAP, Cabral JMS (1994) Continuous lipolysis in a reversed micellar membrane bioreactor. *Bioprocess Eng* 10: 21–27

51 Prazeres DMF, Cabral JMS (1994 b) Enzymatic membrane bioreactors and their applications. *Enzyme Microb Technol* 16: 738–750

Strategies for Improving the Lipase-Catalyzed Preparation of Chiral Compounds

Uwe T. Bornscheuer

Contents

Methods and Tools in Biosciences and Medicine
Methods in non-aqueous enzymology, ed. by M. N. Gupta
© 2000 Birkhäuser Verlag Basel/Switzerland

1   Introduction

1.1   Lipases

Lipases (triacylglycerol-hydrolases, EC 3.1.1.3) occur in plants, mammalians and microorganisms, where their physiological role is believed to be digestive. Common natural substrates are triglycerides and derivatives thereof such as di- and monoglycerides. In contrast to carboxyl esterases (EC 3.1.1.1), they act efficiently on water-insoluble substrates by binding to the water-organic interface, whereas the organic interface usually consists of a triglyceride. This binding not only places the lipase close to the substrate, but also increases the catalytic efficiency drastically by a phenomenon called interfacial activation. This observation was made long time ago by measuring lipase activity at different substrate concentrations. However, only recently this phenomenon could be explained at a molecular level by X-ray crystallographic analysis of lipase crystals containing bound transition state analogs. It was found that in most lipases, a helical lid covers the active site thus blocking access of substrates. Upon binding to a hydrophobic interface such as a lipid droplet, the lid opens enabling substrates access to the active site thus increasing the catalytic activity of lipases. Structures of 11 lipases have been elucidated and it turned out that all of them share a similar motif, called the α/β-hydrolase fold. This consists of a core of eight (mostly parallel) β-sheets surrounded on both sides by α-helices. Lipases are serine esterases and the catalytic machinery is similar to that of serine proteases [1–7].

Probably due to their natural function at the oil-water interface, lipases exhibit an extremely high stability in the presence of organic solvents and are even active in supercritical carbon dioxide. Beside triglycerides, they also accept a wide variety of non-natural substrates, which are usually converted in a highly stereo- and/or regioselective manner. For these reasons, lipases have found applications in various fields of biotransformations. These can be classified according to the nature of the substrates into three main categories: i) modification of fats and oils; ii) acylation/deacylation of carbohydrates and protecting/deprotecting of peptides; and iii) synthesis of chiral compounds. In this article, the major focus will be given to the last application. For modification of fats and oils, reactions involving carbohydrates and peptides and the resolution of carboxylic acids, amines, thiols, lactones etc. readers are referred to a number of reviews [8–20].

In this article, some guidelines for the optimization of reaction conditions with special emphasis on their effect on the enantioselectivity are provided. The influence of water activity is not covered, because this is already discussed in the first chapter. Beside the examples reviewed in this article, very recent work suggests, that the enantioselectivity of hydrolases can also be significantly enhanced by directed evolution of biocatalysts. In the only example published yet, the enantioselectivity of a lipase from *Pseudomonas aeruginosa* towards 2-methyl decanoate was increased from E=1 (wild-type) to E=10 (mutant) by four sequential rounds of directed evolution using error-prone PCR followed by screening based on optically pure *p*-nitrophenol esters [21].

1.2 Choice of solvent system

The choice of a solvent is usually determined by: i) stability and activity of the lipase; ii) stability of substrate or product; iii) solubility of substrate or product; iv) total synthetic strategy, e.g. avoidance of water in subsequent chemical reactions; v) compatibility of solvents with product application, e.g. no toxic or halogenated solvents are allowed for food-stuffs; vi) costs etc. In addition, for the preparation of optically pure compounds, lipases may exhibit different enantioselectivity in different solvent systems.

Lipase-catalyzed reactions have been performed in a variety of solvents and solvent systems including reverse micelles and supercritical media as given in Table 1 [22].

Table 1   Examples of solvent systems for lipase-catalyzed reactions.

Solvent system	Pro	Contra
Water or buffer	high lipase stability, high rate	difficult lipase recovery, no water labile substrates, low substrate, solubility
Water & water-miscible solvent	enhanced substrate solubility, high lipase stability, high rate	difficult lipase recovery, no water, labile substrates
Water & water-immiscible solvent	high lipase stability, high rate, enhanced substrate solubility, good interface, facilitated substrate/product recovery	difficult lipase recovery, no water, labile substrates
Organic solvent (water < 5%)	high lipase stability, easy lipase recovery, good interface, facilitated substrate/product recovery	low rate, heterogeneuous system
Supercritical solvent	high mass transfer, high lipase stability, no solvent removal, facilitated substrate/product recovery	low rate, high equipment costs
Reverse micelles	large surface area, easy photometric studies	low yield, low rate, difficult substrate/product & enzyme recovery

In case of supercritical fluids, most research was performed in supercritical carbon dioxide (SCCO$_2$), because it is non-flammable, non-toxic, cheap, and reaches the supercritical state at low temperature (31.1°C) and moderate pressure (72.9 atm). Its solvating property is comparable to acetone. Supercritical fluids have densities and dissolving powers near those of a liquid, but viscosities near that of a gas. First introduced by Nakamura et al. (1986) [23], several examples have been described and reviewed [8, 22, 24, 25].

Reactions were performed in batch as well as continuous reactor systems wherein recycling of carbon dioxide and separation of products were possible. Most reactions used lipase from *Rhizomucor miehei* (RML) for the modification of triglycerides and derivatives. Only a few examples dealt with the resolution of racemates and in these cases usually the performance of lipases in $SCCO_2$ was compared to its performance in organic solvents. In most cases the enantioselectivity of lipases in $SCCO_2$ is similar to or slightly lower than in organic solvents [26–28]. Ikushima et al. (1995) found that the enantioselectivity of the acetylation of (±)-citronellol in $SCCO_2$ with lipase from *Candida rugosa* (CRL) varied with pressure [29]. On the other hand, Rantakylae and Aaltonen (1994) found no changes in enantioselectivity at different pressures for the RML-catalyzed esterification of ibuprofen with *n*-propanol [30].

1.3   Choice of acyl donor

A wide variety of acyl donors have been used in the lipase-catalyzed preparation of optically pure compounds. The simplest and cheapest acyl donors are acids or simple esters (e.g. methyl, ethyl, glyceryl esters), however acylation with these donors is usually slow due to an unfavorable equilibrium constant. To enhance the reaction rate and achieve maximum conversion, water or alcohol (e.g. methanol, ethanol) generated during a (trans)esterification with these acyl donors can be removed by evaporation [31], distillation [32], or by addition of molecular sieves or inorganic salts [33]. Much more versatile are activated acyl donors as shown in Figure 1. Initially, activated esters were employed, where the alcohol is a better leaving group such as in trifluoroethyl butyrate. Thioesters have the disadvantage of releasing volatile thiols having a very undesired flavor [34, 35]. Oxime esters allow a faster acylation compared to simple and even enol esters, but their use is hampered by the difficulty in removal of the non-volatile oximes and that they are not commercially available [36]. Acid anhydrides also can allow an irreversible acylation of alcohols, but their use in stereoselective reactions is associated with two major disadvantages: i) the concomitant release of the carboxylic acid may decrease the enantioselectivity of the reaction and; ii) lower the pH in the microenvironment of the enzyme, thus leading to inactivation and/or altered activity. The effect on the enantioselectivity might be circumvented by addition of inorganic salt as shown by Berger et al. (1990) (Fig. 2) [37]. An interesting example is the combination of an acylation with a facilitated work-up of the reaction mixture. By using succinic acid anhydride a simple extraction procedure allowed the separation of unreacted alcohol from the charged succinic acid half ester [38].

The most versatile acyl donors are undoubtedly enol esters, such as vinyl acetate or isopropenyl acetate. For both, the product alcohol tautomerizes to a carbonyl compound (acetaldehyde or acetone, respectively) thereby driving the acylation reaction to completion and eliminating potential product inhibi-

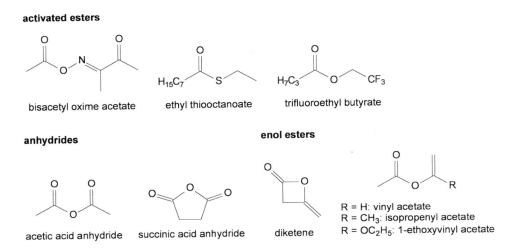

Figure 1   Activated esters, anhydrides and enol esters for lipase-catalyzed acylations.

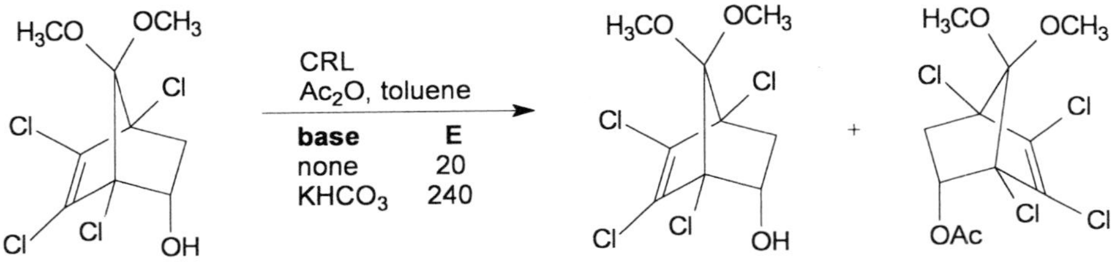

Figure 2   Effect of inorganic salt addition on enantioselectivity (CRL: *Candida rugosa* lipase)

tion. Enol esters were introduced by several groups almost simultaneously [39–44] and became since then the acyl donors of choice in most biotransformations. The only disadvantage of vinyl esters is that some lipases (especially lipases from *Candida rugosa* and *Geotrichum candidum*) do not tolerate the liberated acetaldehyde, which presumably forms a Schiff base with Lysine residues near the active site of lipases [45]. Acetone from isopropenyl acetate might be less harmful. Some reports described a lower enantioselectivity while using vinyl acetate due to the presence of small amounts of water in the lipase preparation leading to the formation of acetic acid. This acid can lead to a reduced enantioselectivity in a competing acylation with vinyl acetate [46]. Working with dry conditions and using an excess of vinyl acetate might circumvent this problem. Recently, diketene was proposed as alternative acylating agent. In this case the cyclic enol ester offers the advantage that no by-products are formed – only the acetoacetate ester of the substrate alcohol is generated – but the enantioselectivities were slightly lower compared to those for vinyl acetate [47–49].

1.4   Influence of immobilization on enantioselectivity and activity

In aqueous solution, lipases are completely dissolved and usually a high activity is observed. In contrast, the enzyme is only suspended as solid particles when working in organic solvents. This facilitates isolation of enzyme by filtration or centrifugation, but is usually associated with a drastically reduced activity of the order of at least one magnitude. The simplest way to increase the activity in organic solvents is the adsorption of the lipase on a carrier with a large surface area, such as Celite. This not only increases the activity but also the stability of the enzyme is in general much higher. Purified lipases denature quickly in organic solvents [50] and as a consequence most commercial lipases are sold in an immobilized form or at least adsorbed on an inorganic support. Banfi et al. (1995) reported a 7–20 times faster reaction for lipase from porcine pancreas adsorbed on Celite [51]. The addition of sugars such as lactose [52, 53] also can increase catalytic activity.

Alternatively, the enzyme can be modified to create an organic solvent soluble lipase [8]. Kikkawa et al. (1989) coupled polyethylene glycol (PEG) to the free amino groups of lipase from *Pseudomonas* sp., which was then soluble in benzene, toluene and chlorinated hydrocarbons. Simple recovery was achieved by precipitating the PEG-lipase by addition of hexane [54]. Similarly, Kodera et al. (1994) used a comb-shaped polymer to dissolve a lipase in organic solvents [55]. A more simple strategy is to coat the lipase with a lipid or surfactant. Okahata and Ijiro (1988) coated lipase from *Rhizopus* sp. with the nonionic amphiphile didodecyl *N*-D-glucono-*L*-glutamate. The modified lipase was soluble in most organic solvents and more than 100 times more active then suspended enzyme [56, 57]. The surfactant resembles a hydrophobic interface and may lead to the opening of the lipase lid [58, 59]. This can also explain why the increase in activity diminishes with time, because the surfactant is only losely bound to the enzyme.

These techniques not only led to an increased activity, but can also affect enantioselectivity after imprinting of the enzyme. Lyophilization of *Candida rugosa* lipase with (*R*)-1-phenylethanol followed by lipid coating increased E from 5.5 to 77 [60].

Alternatively, lipases can be stabilized by covalent immobilization and numerous examples can be found in literature [61, 62]. Only two examples – cross-linked enzyme crystals (CLECs) and sol-gel immobilized lipases – will be discussed here, because a profound effect on enantioselectivity was observed. CLECs are pure crystals of enzymes cross-linked with glutaraldehyde and two lipase CLECs from *Pseudomonas cepacia* (CLEC-PC) and *Candida rugosa* (CLEC-CR) are currently available from Altus Biological Inc. (Cambridge, MA, USA) [63, 64]. Besides higher stability and activity at higher temperatures, CLEC-CR showed an improved enantioselectivity in the resolution of ketoprofen. Enantioselectivity increased from E=5.2 for commercial lipase to E=64 for

CLEC-CR [65, 66]. This was attributed to the removal of a non-selective ester-ase and a conformational change during purification. Lipases entrapped in hy-drophobic sol-gels (prepared by hydrolysis of tetramethoxysilane and alkyltri-methoxysilanes in the presence of lipase) also showed up to 100 fold increased activity [67].

1.5   Kinetic resolution vs asymmetric syntheses

In a highly stereoselective lipase-catalyzed kinetic resolution of a racemic sub-strate only one enantiomer will be converted to the product and at maximum conversion, isolated yield is 50% (Fig. 3). If the stereoselectivity is not high en-ough, only the remaining substrate can be isolated in optically pure form but yields will then be further decreased. Another option to enhance the enantio-meric purity can be isolation of enriched material followed by a second kinetic resolution (also called *recycling*) or by sequential kinetic resolution (also called *in situ* recycling). Several examples for recycling have been published [68, 69]. These sequential kinetic resolutions are especially well-suited for C_2-sym-metric diols [70], however these special cases are outside the scope of this arti-cle, examples can be found in [8].

In contrast to kinetic resolutions, asymmetric syntheses starting from *meso-* or prostereogenic compounds allow the preparation of chiral compounds in up to 100% chemical yield and – at least in theory – the enantiomeric purity of the product remains constant and independent of the conversion (Fig. 3). Thus, it is always advantageous to perform an asymmetric synthesis instead of a ki-netic resolution. Numerous examples can be found in recent reviews [8, 71].

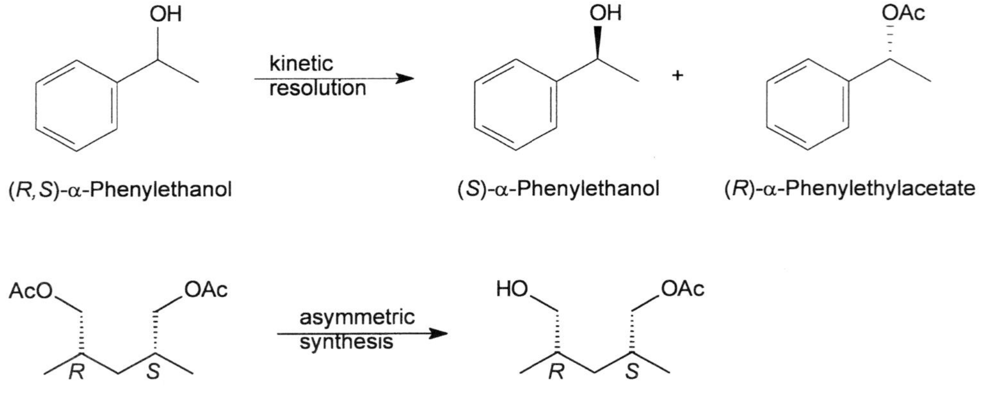

Figure 3   Kinetic resolution vs asymmetric synthesis.

2 Materials

Chemicals
All chemicals were purchased from common suppliers at the highest purity available.

Enzymes
In all protocols given below for the preparation of optically pure compounds, a screening of several commercial enzymes (lipases and pig liver esterase) revealed, that the most reactive and enantioselective lipase was from *Pseudomonas cepacia* (PCL, Amano PS, Amano Pharmaceuticals, Nagoya, Japan). The enzyme was usually used directly as provided by the supplier without any treatment or purification.

Equipment
- Support: Celite and Hyflo Super Cel used for immobilization of PCL were from Fluka
- Sol-gel immobilized PCL was a gift from Prof. Reetz, Mülheim, Germany.

Equations
- Equation 1
Calculation of enantioselectivity E (ee_S: enantiomeric purity of starting material, ee_P: enantiomeric purity of product, *c*: extent of conversion):

$$E = \frac{\left(\frac{V_{max}}{k_M}\right)_R}{\left(\frac{V_{max}}{k_M}\right)_S} = \frac{\ln\left[1 - c\left(1 + ee_P\right)\right]}{\ln\left[1 - c\left(1 + ee_P\right)\right]} = \frac{\ln\left[\left(1 - c\right)\left(1 + ee_S\right)\right]}{\ln\left[\left(1 - c\right)\left(1 + ee_S\right)\right]} = \frac{\ln\left[\frac{1 - ee_S}{1 + \frac{ee_S}{ee_P}}\right]}{\ln\left[\frac{1 - ee_S}{1 + \frac{ee_S}{ee_P}}\right]}$$

- Equation 2
Calculation of enantiomeric excess by molar fraction of enantiomers A and B:

$$\%ee = \frac{x_A - x_B}{x_A - x_B}$$

Calculation of enantioselectivity
Optical purities are usually expressed in % enantiomeric excess (% ee), which can be calculated from the molar fraction (or by their peak areas from e.g. gas chromatographic analysis on a chiral column) of each enantiomer using equation 2. However, the enantiomeric purity of the reactants may vary as the reaction proceeds and a comparison of two kinetic resolutions only makes sense at the same extent of conversion (see section 4). A convenient way to circumvent

this problem is by comparing the inherent enantioselectivity (or enantiomeric ratio), E, which was first introduced by Charles Sih's group [72]. An E-value of 1 resembles a non-selective reaction and an E-value >100 resembles a highly stereoselective reaction. In practice, reactions with E >20 are usually sufficient. The E-values given in the Protocols section were calculated using equation 1. A simple program to calculate E is freely available at http://www-orgc.-tu-graz.ac.at [73].

3   Methods

3.1   Immobilization of lipase

For immobilization, 1 g lipase powder was dissolved in 8 ml phosphate buffer (pH 7.5) and centrifuged. The supernatant was mixed with 2 g Celite or Hyflo Super Cel and the mixture was sucked through a Büchner funnel. The solid was washed several times with chilled acetone and first dried in air and finally in a desiccator under vacuum.

3.2   Assay for lipase activity

Lipase activity was determined using a pH-stat assay system (Metrohm, Herisau, Schweiz). 20 mL assay solution (5% (w/v) olive oil, 2% (w/v) gum arabicum, mixed in an ultraturrax for 5 min) and 470 µL $CaCl_2$ solution (22% (w/v)) were incubated at 37 °C. A known amount of lipase was dissolved in phosphate buffer, centrifuged and an aliquot of the supernatant was assayed. Released fatty acids were titrated automatically with 0.1 N NaOH. One unit (U) of activity is defined as the amount of lipase that liberates 1 µmol of fatty acid from olive oil in 1 min at 37 °C.

3.3   Protocols

Protocol 1   Resolution of protected glycerol derivatives

The kinetic resolution of 1,2-*O*-protected glycerols in organic solvents using vinyl esters as acyl donors was studied (section 5.1). Because lipases show very low enantioselectivity towards primary alcohols, the nature of the protecting group was varied with the intention to increase the discrimination of PCL between both enantiomers.

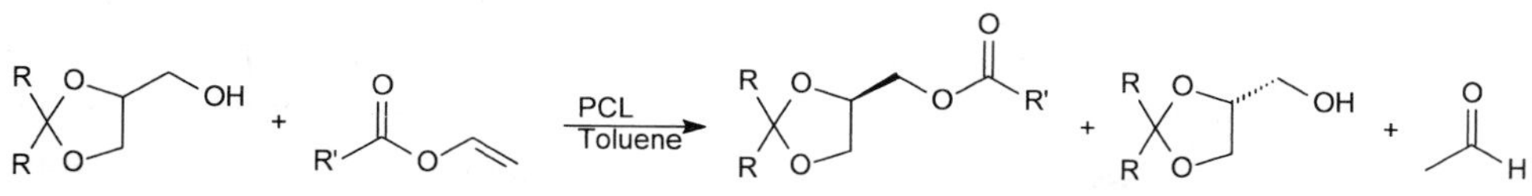

Figure 4 Lipase-catalyzed resolution of protected glycerol derivatives 1–8 [74]
For residues R see Table 2.

Table 2 Results of lipase-catalyzed resolution of protected glycerol derivatives 1–8 [74]. The reaction scheme is shown in Figure 4.

No.	R	Vinyl acetate (R′=CH$_3$)		Vinyl butyrate (R′=C$_3$H$_7$)	
		% ee$_S$	E	% ee$_S$	E
1	Methyl-	78	2.3	42	4.0
2	Ethyl-	63	1.7	35	1.7
3	n-Propyl-	51	5.5	29	1.4
4	i-Propyl-	79	5.5	46	8.5
5	neo-Butyl-	29	1.3	15	1.1
6	-(CH$_2$)$_4$-	7	1.0	7	1.3
7	-(CH$_2$)$_5$-	35	1.3	37	2.0
8	-(CH$_2$)$_6$-	20	1.6	45	n.d.

With the exception of commercially available isopropylidene glycerol 1 (IPG), all 1,2-*O*-protected glycerols 2–8 (Fig. 4 and Tab. 2) were synthesized from glycerol (130 mmol) and the corresponding ketone (100 mmol) in toluene (100 mL) using *p*-toluene sulfonic acid (1 mmol) as catalyst. Synthesis was performed in a two-necked round bottom flask equipped with a Dean-Stark trap and a condenser. The mixture was refluxed until formation of water stopped. This was controlled by the amount of water collected in the Dean-Stark trap. The reaction mixture was washed with ice water (30 mL), saturated Na$_2$CO$_3$ solution (30 mL) and water (30 mL). After drying over Na$_2$SO$_4$ and evaporation of the solvent the crude product was purified by silica gel column chromatography (petrol ether:diethyl ether, 2:1) and obtained as colorless liquids in yields ranging from 60–80%. Structures and purity were confirmed by NMR and mass spectra.

Transesterifications were performed in 10 ml glass-stoppered round bottom flasks thermostated by an oil bath at 40 °C. In a typical experiment 1.5 mmol 1,2-*O*-protected glycerol and 1.5 mmol vinyl ester were dissolved in 3 mL toluene and 50 units of lipase (PCL) were added. Samples withdrawn from the reaction mixture were centrifugated. An aliquot of 30 μL was dissolved in 200 μL dichloromethane in an Eppendorf test tube and 50 μL trifluoroacetic acid anhydride (TFAA) was added for derivatization to facilitate separation of enantiomers in gas chromatographic analysis. After 10 min solvent and unreacted TFAA were evaporated using nitrogen, a small aliquot of dichloromethane was added and the samples were analyzed by GC. Under these conditions, no racemization was observed, as determined in the derivatization of pure (*R*)-IPG with TFAA and subsequent analysis. GC analysis was performed on a gas chromatograph (HRGC Mega 2 series, Autosampler A 200S, software package Chrom Card for Windows, Fisons Instruments, Mainz-Kastel, Germany) using a chiral stationary phase (hydrodex β-3P = heptakis-(2,3,6-tri-*O*-methyl)-β-cyclodextrin (25 m, x 0.25 mm, Macherey & Nagel, Düren, Germany) and a flame ionization detector. Temperature programs for the analysis of each compound have been published [74]. The conversion was calculated from the enantiomeric excess of substrate and product as described in section 2 and reference [72].

Protocol 2   Desymmetrization of *meso*-diacetates

As already outlined in section 1.5, the desymmetrization of *meso*-compounds allows the preparation of optically pure compounds in yields up to 100%. In this study, *meso*-diacetates were subjected to lipase-catalyzed hydrolysis and again the influence of substrate structure on enantioselectivity was of special interest (section 5.1).

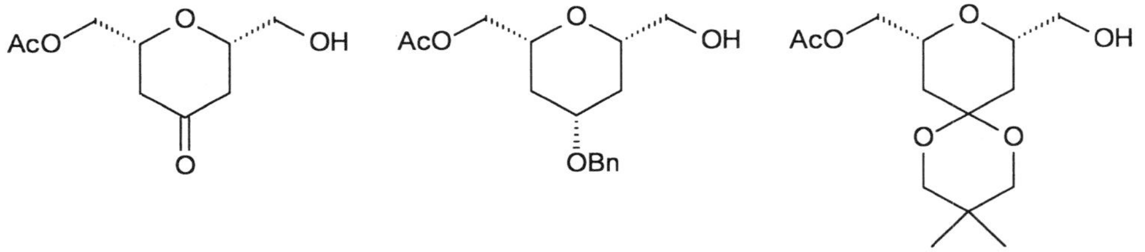

9a   **9b**   **9c**

Figure 5   Lipase-catalyzed desymmetrization of *meso*-diacetates afforded **9a–c** [75].

Meso-diacetates of **9a–c** (Fig. 5) (0.5 to 3.5 mmol) were dissolved in the chosen solvent (4 mL) and 0.5 molar phosphate buffer solution (16 mL) was added. The biphasic reaction mixture was vigorously stirred using a magnetic stirrer and PCL (1500 units/mmol) was added. Reactions were performed at room temperature and pH was maintained at 7 by automatic addition of 1 N NaOH (pH-stat) until desired conversion was achieved as calculated from NaOH consumption (approx. 22 hrs). The reaction was then terminated by addition of ethanol/acetone (1:1). The mixture was extracted six times with ether. The combined organic layers were washed with brine, dried (MgSO₄) and concentrated *in vacuo* to afford a crude mixture that was purified by silica gel column chromatography (petrol ether:diethyl ether, 1:3) to yield the monoacetates (-)-**9a-c** in yields of 80–88%. Enantiomeric excess and absolute configuration was determined by transforming benzylether (-)-**9b** or spiroketal (-)-**9c** into ketone

(-)-9a, followed by ^{1}H-NMR spectroscopy in the presence of (+)-Eu(hfc)$_3$. In case of (-)-9b in addition the Mosher ester was prepared from S-MTPA-Cl. Absolute configuration was further confirmed by X-ray analysis of the p-bromobenzoyl derivative of spiroketal (-)-9c [75].

Protocol 3   Resolution of 3-hydroxy esters in organic solvents

The 3-hydroxy esters 10a–c (Fig. 6) used in this work are bifunctional molecules. Lipase-catalyzed resolution was either performed as transesterification at the hydroxy group using vinyl esters or as hydrolysis of the ester function (section 5.2). Thus, a greater flexibility in terms of reaction conditions and choice of acyl donors is feasible.

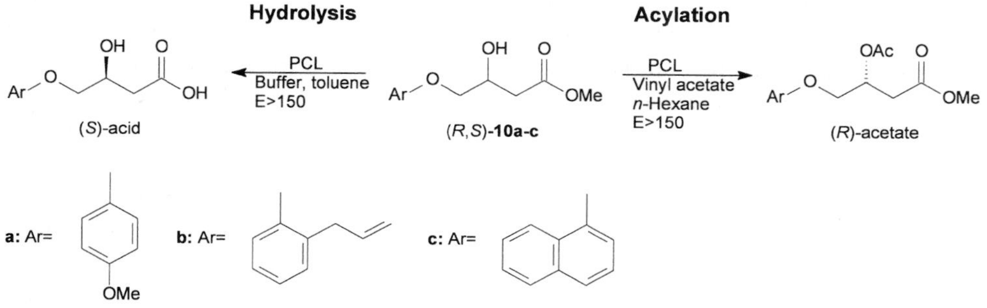

Figure 6   Resolution of 3-hydroxyesters 10a–c [82] as examples for secondary alcohols.

Transesterifications of 3-hydroxy esters 10a–c were carried out at 37 °C in 10 mL glass stoppered round bottom flasks in an oil bath. The magnetic stirrer speed was kept at 400 rpm. In a typical experiment 0.3 mmol (R,S)-10a–c, 0.3 mmol vinyl acetate, 50 mg PCL and 2 mL hexane were added. Samples withdrawn during the reaction were centrifugated to separate the enzyme. The supernatant was derivatized with trifluoroacetic acid anhydride as described in Protocol 1 and analyzed by GC. Hydrolysis was performed in a biphasic reaction system consisting of 50 mM phosphate buffer (pH 7):toluene (1:1) at 37 °C in a pH-stat system (Metrohm, Buchs, Switzerland). After consumption of 0.1 M NaOH indicated approximately 50% conversion, the reaction was stopped. After separation of substrate and product (using silica gel chromatography), the acid was converted into the methyl ester by addition of etheral diazomethane solution and analyzed by GC using the same column and equipment as described in Protocol 1.

Protocol 4   On-line measurement of the conversion in lipase-catalyzed kinetic resolutions in supercritical carbon dioxide

The intention of this work was to show that lipase-catalyzed resolutions can be performed in supercritical carbon dioxide. Furthermore, a method was developed to monitor the conversion under high-pressure conditions without taking samples (section 5.2). The example described is the kinetic resolution of *(R,S)*-3-Hydroxyoctanoic acid methyl ester (**11**, Figure 7).

PCL was used in its crude form. *(R,S)*-3-Hydroxyoctanoic acid methyl ester (**11**, Figure 7) was synthesized as described [76]. Carbon dioxide (purity 4.5) was obtained from Linde, Hölzriegelskreuth, Germany. The reaction system for kinetic resolutions in supercritical carbon dioxide consisted of a batch reactor (18 mL internal volume, equipped with a borosilicate window for visual control of the phases), an external sample loop (3 mL volume) and a high-pressure flow through cell connected to a UV-detector. All equipment except the UV-detector were placed in a thermostated oven. The reactor containing substrate *(R,S)*-**11** (200 µL) and molecular sieve (10 g) was filled with CO_2 using a high pressure pump (Milton-Roy, Obertshausen, Germany, model minipump duplex NSI 33 R) until 110 atm were reached. The lipase (200 mg) was filled into a separate fixed bed reactor (3 mL column with frits at the ends). After solubilization of *(R,S)*-**11** in $SCCO_2$, vinyl acetate (200 µL) was pumped into the reactor to start the reaction. On-line determination of conversion and control of solubility of the substrates was achieved by circulating the reaction mixture with a magnetic coupled pump (Verder, Düsseldorf, Germany, model V 150.12) and measuring the absorbance via the UV-detector. Off-line sampling was performed by recirculation of the reaction mixture through the external loop and decompressurization of the content into *n*-hexane.

On-line determination of the conversion was based on the increase in absorbance at 320 nm due to acetaldehyde and data was monitored on a recorder. For calibration, off-line samples from the reaction mixture were derivatized with trifluoroacetic acid anhydride and *(R,S)*-**11** was analyzed by gas chromatography using a chiral stationary phase (Lipodex E, Macherey & Nagel, Düren, Germany) as described previously [77]. Conversion was then calculated from enantiomeric excess according to [72]. Solubility of substrates was confirmed by recording the UV-spectra from 180 to 340 nm at different pressures.

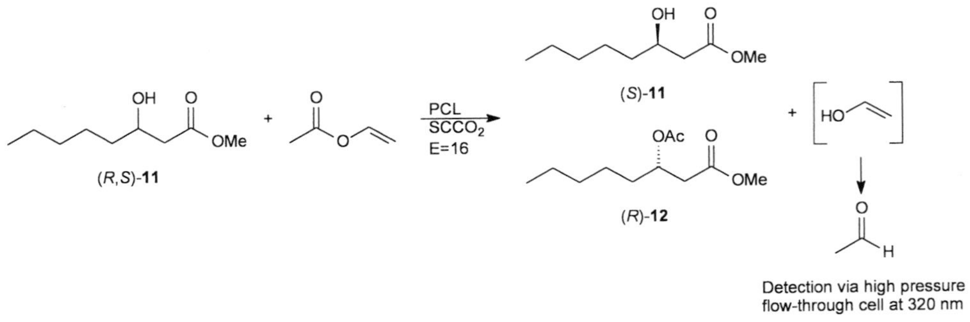

Detection via high pressure
flow-through cell at 320 nm

Figure 7   On-line determination of the conversion by measurement of the absorbance of acetaldehyde generated in the PCL-catalyzed resolution of 3-hydroxyester **11** [76]

4 Troubleshooting

- *Insufficient enantioselectivity*

In all examples – except the protected glycerol derivatives **1–8** – shown in this article, the enantioselectivity of lipase from *Pseudomonas cepacia* was from good to excellent. However, if this is not the case, some further experiments might help to increase the E value. It is recommended to perform optimizations on a small scale (<10 mg substrate can be enough) and to first check several lipases and solvents. Further, it is suggested to perform both reactions, hydrolysis and acylation, because they often differ in enantioselectivity. The latter approach provides more flexibility, because also different acyl donors can be investigated. It might also be useful to lower the reaction temperature even below 0 °C to reduce the conversion of the slow-reacting enantiomer thus enhancing E. From my own experience, immobilization as well as water activity only have a minor effect on enantioselectivity. Depending on whether optically pure product or substrate are of interest, a kinetic resolution should be performed either until ca. 40% (high % ee_P) or ca. 60% conversion (high % ee_S).

- *Problems in calculating enantioselectivity*

In a few cases, ambigious results have been reported when the enantioselectivity E was calculated using Equation 1. This can be explained by the fact that Equation 1 is only valid under the assumption of an irreversible reaction, one substrate/one product and absence of product inhibition. For reverse reactions, the equilibrium constant K has to be taken into account too [83]. In addition, biocatalysts are often mixtures of enzymes of different stereoselectivity and/or stability (e.g. isoenzymes), thus the calculated E reflects a weighted average of all the enzymes. Some of these problems can be eliminated by using a computer program to include product inhibition [84–87] as well as to determine both K and E by fitting a series of ee_S and ee_P measurements (available at http://bendik.mnfak.unit.no).

5 Remarks and Conclusions

5.1 Resolutions of primary alcohols via 'Substrate-Engineering'

In contrast to secondary alcohols (see section 5.2), no reliable rule exists to predict the enantioselectivity for primary alcohols. Compared to secondary alcohols the enantiopreference is often opposite and in most cases the E-values are significantly lower. This was especially true for those primary alcohols bearing an oxygen next to the stereocenter, which is the case for all substrates investigated here. The lower enantioselectivity was related by Weissfloch and Kazlauskas (1995) to the additional CH_2-group introducing a kink between the chiral center and the hydroxy function in primary alcohols [78] (Fig. 8b).

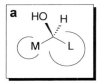

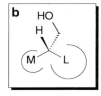

Figure 8 Empirical rules showing the enantiopreference of PCL toward secondary (a) and primary alcohols (b) [78].

The shown enantiomer reacts faster in acylation reactions, in hydrolysis reactions the corresponding ester reacts faster. Please note that PCL shows opposite enantiopreference for these two classes of alcohols and that the rule is not valid for primary alcohols bearing an oxygen next to the stereocenter. (M: medium-sized substituent, L: large substituent).

Our approach to enhance the enantioselectivity of lipases towards primary alcohols was based on variations of the substrate structure. Because variation of reaction conditions is often referred to as medium-engineering, we would like to name this approach substrate-engineering.

Indeed, we found that variation of the substrate structure affects enantioselectivity during our studies on the resolution of glycerol derivatives [74] (Protocol 1) and the desymmetrization of *meso*-diols [75] (Protocol 2).

In the case of protected glycerol derivatives **1–8** (Fig. 4 and Tab. 2) enantioselectivity was increased from e.g. E=2.3 for isopropylidene glycerol **1** to E=5.5 for 2,2-diisopropyl-1,3-dioxolane-4-methanol **4** using vinyl acetate and to E=8.5 using vinyl butyrate [74]. From the data given in Table 2, the following conclusions can be made: i) an increase in the linear chain of the protecting group led to a decrease in the enantiomeric excess of the substrate in the order R' = methyl > ethyl > *n*-propyl; ii) an additional methyl branch influenced the enantiomeric excess in a non-consistant manner (% ee for *i*-propyl >> *neo*-butyl) and; iii) cyclic protecting groups seem to be the worst choice. In all examples, the E values are still not high enough for a preparative synthesis of optically pure compounds.

Beside variation of substrate structure, immobilization also affected enantioselectivity to some extent. PCL was immobilized by adsorption on Celite or Hyflo Super Cel and by using a sol-gel preparation and the acylation of IPG **1** with vinyl butyrate in toluene was studied. The enantioselectivity increased from E=4 (crude, non-immobilized PCL) to E=5.2 (Hyflo Super Cel-PCL), E=7.3 (Celite-PCL) and E=7.5 (Sol-gel-PCL). Using vinyl acetate as acyl donor, almost no change in enantioselectivity was observed [79].

In sharp contrast to the resolution of glycerol derivatives **1–8**, the strategy of substrate-engineering was very successful in the desymmetrization of *meso*-substrates **9a–c** (Fig. 5). These 2,4,6-trifunctionalized *C*-glycosides are versatile building blocks in the synthesis of e.g. Bryostatin 1 [80].

Hydrolysis of the *meso*-diacetate of **9a** revealed that PCL gave best results, when the reaction was performed in a biphasic mixture of phosphate buffer:toluene (1:4). However, maximum enantiomeric excess was only 70% ee for the monoacetate isolated in 70% yield. A substantial increase in enantioselectivity was achieved by using either the diacetate of benzylether **9b** or that of spiroketal **9c** as substrates. In both cases the monoacetates obtained by lipase-catalyzed hydrolysis had an enantiomeric excess > 98% ee and were isolated in 88% yield. Thus, the enantioselectivity of PCL increased, although variation of the substrate structure occured far away from the ester group. We proposed that this change in enantioselectivity was due to a reduced flexibility in the conformation of benzylether **9b** and spiroketal **9c**, because all three substituents at the tetrahydropyrane ring adopt an equatorial position [80].

5.2   Kinetic resolutions of secondary alcohols

Based on a detailed literature study, Kazlauskas et al. (1991) proposed a model to predict which enantiomer of a secondary alcohol reacts faster in lipase-catalyzed reactions. This empirical rule is based on the size of the substituents at the chiral center. High enantioselectivity is observed when one substituent is a medium size residue (e.g. a methyl group) and the other substituent a large residue (e.g. an aromatic ring). The model (Fig. 8a) was valid for the vast majority of compounds in reactions catalyzed by lipases from *Pseudomonas cepacia* or *Candida rugosa* [81].

We have investigated the resolution of several 3-hydroxyesters in organic solvents as well as in supercritical carbon dioxide (see below and Protocols 3–4). The arylaliphatic 3-hydroxyesters **10a–c** (Fig. 6) are direct precursors of β-blockers, e.g. **10b** for Alprenolol and **10c** for Propranolol [82]. Biotransformations were performed as acylations using enol esters such as vinyl acetate or hydrolysis in biphasic mixtures. The optimization of reaction conditions, e.g. solvent, water activity, lipase purity and acyl donor, has been already described in a previous paper for several aliphatic 3-hydroxyester, such as **11** [77].

A comparison of the enantioselectivities found in the conversion of arylaliphatic 3-hydroxyesters **10a–c** with the reactions using aliphatic 3-hydroxyester **11** revealed that the introduction of the aromatic groups indeed led to a significantly increased enantioselectivity from E=16 (**11**) to E>150 (**10a–c**).

The resolution of 3-hydroxy esters was also investigated in supercritical carbon dioxide (SCCO$_2$) using the reactor device described in Protocol 4. Reactions are best performed at 40 °C, 110 atm with PCL and vinyl acetate as acyl donor (Fig. 7). Enantioselectivities were similar compared to reactions in organic solvents, but reaction rates were 5-fold lower in SCCO$_2$ [26]. The progress of the reaction could be monitored by taking samples from a built-in sampling loop, however this was accompanied with a pressure drop and might cause a disturbance of the reaction system. In order to circumvent this problem, we monitored the formation of the by-product acetaldehyde generated from vinyl acetate in stoechiometric amounts in the lipase-catalyzed acylation of 3-hydroxy ester **11** through a high-pressure flow-through cell at 320 nm. The on-line data agreed very well with off-line values up to 60% conversion [76] (see Protocol 4), which is sufficient for an enantioselective kinetic resolution of racemates.

Acknowledgments

I would like to thank all my co-workers, who contributed to the examples presented in protocols 1–4 of this article. Furthermore, I am especially grateful to Prof. Romas J. Kazlauskas (Dept. of Chemistry, McGill University, Quebec, Canada) for many useful discussions during the preparation of our hydrolase book [8].

Abbreviations

(+)-Eu(hfc)$_3$ Tris-[3-heptafluoropropyl-hydroxymethylen)-d-camphorato]-
                europium (III)

S-MTPA-Cl   S(-)-α-methoxy-alpha-trifluoromethylphenyl acetic acid chloride

References

1   Cambillau C, van Tilbeurgh H (1993) Structure of hydrolases - lipases and cellulases. *Curr Opin Struct Biol* 3: 885–895

2   Cygler M, Schrag JD, Ergan F (1992) Advances in structural understanding of lipases. *Biotechnol Genet Eng Rev* 10: 143–184

3   Derewenda U, Brzozowski AM, Lawson DM, Derewenda ZS (1992) Catalysis at the interface: the anatomy of a conformational change in a triglyceride lipase. *Biochemistry* 31: 1532–1541

4   Derewenda U, Swenson L, Green R, Wei Y, Yamaguchi S, Joerger R, Haas MJ, Derewenda ZS (1994) Current progress in crystallographic studies of new lipases from filamentous fungi. *Protein Eng* 7: 551–557

5   Dodson GG, Lawson DM, Winkler FK (1992) Structural and evolutionary relationships in lipase mechanism and activation. *Faraday Discuss* 95–105

6   Ollis DL, Cheah E, Cygler M, Dijkstra B, Frolow F, Franken SM, Harel M, Remington SJ, Silman I, Schrag J et al. (1992) The α?β hydrolase fold. *Protein Eng* 5: 197–211

7   Ransac S, Carriére F, Rogalska E, Verger R, Marguet F, Buono G, Melo EP, Cabral JMS, Egloff MPE, van Tilbeurgh H et al. (1996) The kinetics specificities and structural features of lipases in: Molecular Dynamics of Biomembranes. In: op den Kamp, J.Book (eds): *NATO ASI series* H 96. Springer, Berlin, 265–304

8   Bornscheuer UT, Kazlauskas RJ (1999) *Hydrolases in organic synthesis - regio- and stereoselective biotransformations.* VCH-Wiley, Weinheim

9   Villeneuve P, Foglia TA (1997) Lipase specificities: Potential applications in lipid bioconversions. *Inform* 8: 640–650

10   Bornscheuer U (1995) Lipase-catalyzed syntheses of monoacylglycerols. *Enzyme Microb Technol* 17: 578–586

11   Marangoni AG, Rousseau D (1995) Engineering triacylglycerols: The role of interesterification. *Trends Food Sci Technol* 6: 329–335

12   Haas MJ, Joerger RD (1995) Lipases of the genera *Rhizopus* and *Rhizomucor*: versatile catalysts in nature and the laboratory. In: Khachatourians, G. G. and Hui, H. Y.Book (eds): *Food Biotechnology: Microorganisms.* VCH, Weinheim, 549–588

13   Vulfson EN (1994) Industrial applications of lipases. In: Wooley, P. and Peterson, S. B.Book (eds):*Lipases their structure biochemistry and application* Cambridge Universiy Press, Cambridge, 271–288

14   Adlercreutz P (1994) Enzyme-catalysed lipid modification. *Biotechnol Gen Eng Rev* 12: 231–254

15   Björkling F, Godtfredsen SE, Kirk O (1991) The future impact of industrial lipases. *Trends Biotechnol* 9: 360–363

16   Waldmann H, Sebastian D (1994) Enzymic protecting group techniques. *Chem Rev* 94: 911–937

17 Bashir NB, Phythian SJ, Roberts SM (1995) Enzymatic regioselective acylation and deacylation of carbohydrates. In: Roberts, S. M.Book (eds):*Preparative Biotransformations* Wiley, 0:0.43–0:0.74.

18 Thiem J (1995) Applications of enzymes in synthetic carbohydrate chemistry. *FEMS Microbiol Rev* 16: 193–211

19 Wong C-H (1995) Enzymatic and chemo-enzymatic synthesis of carbohydrates. *Pure Appl Chem* 67: 1609–1616

20 Riva S (1996) Regioselectivity of hydrolases in organic media. In: Koskinen, A. M. P. and Klibanov, A. M.Book (eds):*Enzymatic Reactions in Organic Media* Chapman & Hall, Glasgow, 140–169

21 Reetz MT, Zonta A, Schimossek K, Liebeton K, Jaeger K-E (1997) Erzeugung enantioselektiver Biokatalysatoren für die Organische Chemie durch In-vitro-Evolution. *Angew Chem* 109: 2961–2963

22 Ballesteros A, Bornscheuer U, Capewell A, Combes D, Condorét J-S, König K, Kolisis FN, Marty A, Menge U, Scheper U et al. (1995) Enzymes in non-conventional media. *Biocatal Biotransform* 13: 1–42

23 Nakamura K, Chi YM, Yamada Y, Yano T (1986) Lipase activity and stability in supercritical carbon dioxide. *Chem Eng Commun* 45: 207–212

24 Aaltonen O, Rantakylae M (1992) Biocatalysis in supercritical CO_2. *Chemtech* 21: 240–248

25 Nakamura K (1990) Kinetics of lipase-catalyzed esterification in supercritical CO_2. *Trends Biotechnol* 8: 288–291

26 Capewell A, Wendel V, Bornscheuer U, Meyer HH, Scheper T (1996) Lipase-catalyzed kinetic resolution of 3-hydroxy esters in organic solvents and supercritical carbon dioxide. *Enzyme Microb Technol* 19: 181–186

27 Martins JF, Sampaio TC, Carvalho IB, da Ponte MN, Barreiros S. (1992) in High Press. Biotechnol., vol. 224 (Balny, C., Hayashi, R., Heremans, K. and Masson, P., eds.), pp. 411–415

28 Michor H, Marr R, Gamse T, Schilling T, Klingsbichel E, Schwab H (1996) Enzymic catalysis in supercritical carbon dioxide: comparison of different lipases and a novel esterase. *Biotechnol Lett* 18: 79–84

29 Ikushima Y, Saito N, Arai M, Blanch HW (1995) Activation of a lipase triggered by interactions with supercritical carbon dioxide in the near critical region. *J Phys Chem* 99: 8941–8944

30 Rantakylae M, Aaltonen O (1994) Enantioselective esterification of ibuprofen in supercritical carbon dioxide by immobilized lipase. *Biotechnol Lett* 16: 825–830

31 Björkling F, Godtfredsen SE, Kirk O (1989) A highly selective enzyme-catalyzed esterification of simple glucosides. *J Chem Soc, Chem Commun* 934–935

32 Bloomer S, Adlercreutz P, Mattiasson B (1992) Facile synthesis of fatty acid esters in high yields. *Enzyme Microb Technol* 14: 546–552

33 Kvittingen L, Sjursnes BJ, Anthonsen T, Halling P (1992) Use of salt hydrates to buffer optimal water level during lipase catalyzed synthesis in organic media: a practical procedure for organic chemists. *Tetrahedron* 48: 2793–2802

34 Öhrner N, Martinelle M, Mattson A, Norin T, Hult K (1992) Displacement of the equilibrium in lipase catalyzed transesterification in ethyl octanoate by continuous evaporation of ethanol. *Biotechnol Lett* 14: 263–268

35 Frykman H, Öhrner N, Norin T, Hult K (1993) S-Ethyl thiooctanoate as acyl donor in lipase-catalyzed resolution of secondary alcohols. *Tetrahedron Lett* 34: 1367–1370

36 Ghogare A, Kumar GS (1990) Novel route to chiral polymers involving biocatalytic transesterification of O-acryloyl oximes. *J Chem Soc, Chem Commun* 134–135

37 Berger B, Rabiller CG, Königsberger K, Faber K, Griengl H (1990) Enzymatic acylation using acid anhydrides: crucial removal of acid. *Tetrahedron: Asymmetry* 1: 541–546

38 Terao Y, Tsuji K, Murata M, Achiwa K, Nishio T, Watanabe N, Seto K (1989) Facile process for enzymic resolution of racemic alcohols. *Chem Pharm Bull* 37: 1653–1655

39 Sweers HM, Wong C-H (1986) Enzyme-catalyzed regioselective deacylation of protected sugars in carbohydrate synthesis. *J Am Chem Soc* 108: 6421–6422

40 Degueil-Castaing M, de Jeso B, Drouillard S, Maillard B (1987) Enzymatic reactions in organic synthesis. 2. Ester interchange of vinyl esters. *Tetrahedron Lett* 28: 953–954

41 Wang YF, Wong C-H (1988) Lipase-catalyzed irreversible transesterification for preparative synthesis of chiral glycerol derivatives. *J Org Chem* 53: 3127–3129

42 Wang YF, Lalonde JJ, Momongan M, Bergbreiter DE, Wong C-H (1988) Lipase-catalyzed irreversible transesterifications using enol esters as acylating re-

agents: preparative enantio- and regio-selective syntheses of alcohols glycerol derivatives sugars and organometallics. *J Am Chem Soc* 110: 7200–7205

43 Laumen K, Breitgoff D, Schneider MP (1988) Enzymic preparation of enantio-merically pure secondary alcohols. Ester synthesis by irreversible acyl transfer using a highly selective ester hydrolase from *Pseudomonas* sp. an attractive alternative to ester hydrolyis. *J Chem Soc, Chem Commun* 1459–1461

44 Terao Y, Murata M, Achiwa K (1988) Highly-efficient lipase-catalyzed asymmetric synthesis of chiral glycerol derivatives leading to practical synthesis of *(S)*-propranolol. *Tetrahedron Lett* 29: 5173–5176

45 Weber HK, Stecher H, Faber K (1995) Sensitivity of microbial lipases to acetaldehyde formed by acyl-transfer reactions from vinyl esters. *Biotechnol Lett* 17: 803–808

46 Lundh H, Nordin O, Hedenström E, Högberg HE (1995) Enzyme catalysed irreversible transesterifications with vinyl acetate - are they really irreversible? *Tetrahedron: Asymmetry* 6: 2237–2244

47 Balkenhohl F, Hauer B, Ladner W, Schnell U, Pressler U, Staudenmaier HR. (1993) in Chem. Abstr. 122: 212267., German Patent DE 4329293, Germany

48 Jeromin GE, Welsch V (1995) Diketene a new esterification reagent in the enzyme-aided synthesis of chiral alcohols and chiral acetoacetic acid esters. *Tetrahedron Lett* 36: 6663–6664

49 Suginaka K, Hayashi Y, Yamamoto Y (1996) Highly selective resolution of secondary alcohols and acetoacetates with lipases and diketenes in organic media. *Tetrahedron: Asymmetry* 7: 1153–1158

50 Bornscheuer U, Reif O-W, Lausch R, Freitag R, Scheper T, Kolisis FN, Menge U (1994) Lipase of *Pseudomonas cepacia* for biotechnological purposes: purification crystallization and characterization. *Biochim Biophys Acta* 1201: 55–60

51 Banfi L, Guanti G, Riva R (1995) On the optimization of pig pancreatic lipase catalyzed monoacetylation of prochiral diols. *Tetrahedron: Asymmetry* 6: 1345–1356

52 Sanchez-Montero JM, Hamon V, Thomas D, Legoy MD (1991) Modulation of lipase hydrolysis and synthesis reactions using carbohydrates. *Biochim Biophys Acta* 1078: 345–350

53 Dabulis K, Klibanov AM (1993) Dramatic enhancement of enzymic activity in organic solvents by lyoprotectants. *Biotechnol Bioeng* 41: 566–571

54 Kikkawa S, Takahashi K, Katada T, Inada Y (1989) Esterification of chiral secondary alcohols with fatty acid in organic solvents by polyethylene glycol-modified lipase. *Biochem Int* 19: 1125–1131

55 Kodera Y, Nishimura H, Matsushima A, Hiroto M, Inada Y (1994) Lipase made active in hydrophobic media by coupling with polyethylene glycol. *J Am Oil Chem Soc* 71: 335–338

56 Okahata Y, Ijiro K (1988) A lipid-coated lipase as a new catalyst for triglyceride synthesis in organic solvents. *J Chem Soc, Chem Commun* 1392–1394

57 Okahata Y, Ijiro K (1992) Preparation of a lipid-coated lipase and catalysis of glyceride ester syntheses in homogeneous organic solvents. *Bull Chem Soc Jpn* 65: 2411–2420

58 Mingarro I, Abad C, Braco L (1995) Interfacial activation-based molecular bioimprinting of lipolytic enzymes. *Proc Natl Acad Sci USA* 92: 3308–3312

59 Mingarro I, Gonzalez-Navarro H, Braco L (1996) Trapping of different lipase conformers in water-restricted environments. *Biochemistry* 35: 9935–9944

60 Okahata Y, Hatano A, Ijiro K (1995) Enhancing enantioselectivity of a lipid-coated lipase via imprinting methods for esterification in organic solvents. *Tetrahedron: Asymmetry* 6: 1311–1322

61 Akita H (1996) Recent advances in the use of immobilized lipases directed toward the asymmetric syntheses of complex molecules. *Biocatal Biotransform* 13: 141–156

62 Balcao VM, Paiva AL, Malcata FX (1996) Bioreactors with immobilized lipases: state of the art. *Enzyme Microb Technol* 18: 392–416

63 St. Clair NL, Navia MA (1992) Cross-linked enzyme crystals as robust biocatalysts. *J Am Chem Soc* 114: 7314–7316

64 Margolin AL (1996) Novel crystalline catalysts. *Trends Biotechnol* 14: 223–230

65 Lalonde JJ, Govardhan C, Khalaf N, Martinez AG, Visuri K, Margolin AL (1995) Cross-linked crystals of *Candida rugosa* lipase: highly efficient catalysts for the resolution of chiral esters. *J Am Chem Soc* 117: 6845–6852

66 Persichetti RA, Lalonde JJ, Govardhan CP, Khalaf NK, Margolin AL (1996) *Candida rugosa* lipase - enantioselectivity enhancements in organic solvents. *Tetrahedron Lett* 37: 6507–6510

67 Reetz M, Zonta A, Simpelkamp J (1995) Efficient heterogeneous biocatalysts by entrapment of lipases in hydrophobic sol-gel materials. *Angew Chem Int Ed Engl* 34: 301–303

68 Brown SM, Davies SG, de Sousa JAA (1993) Kinetic resolution strategies II: enhanced enantiomeric excesses and yields for the faster reacting enantiomer in lipase mediated kinetic resolutions. *Tetrahedron: Asymmetry* 4: 813–822

69 Vänttinen E, Kanerva LT (1997) Optimized double kinetic resolution for the preparation of (*S*)-solketal. *Tetrahedron: Asymmetry* 8: 923–933

70 Caron G, Kazlauskas RJ (1991) An optimized sequential kinetic resolution of trans-1,2-cyclohexanediol. *J Org Chem* 56: 7251–7256

71 Schoffers E, Golebiowski A, Johnson CR (1996) Enantioselective synthesis through enzymatic asymmetrization. *Tetrahedron* 52: 3769–3826

72 Chen CS, Fujimoto Y, Girdaukas G, Sih CJ (1982) Quantitative analyses of biochemical kinetic resolutions of enantiomers. *J Am Chem Soc* 104: 7294–7299

73 Kroutil E, Kleewein A, Faber K (1997) A computer program for analysis simulation and optimization of asymmetric catalytic processes proceeding through two consecutive steps. Type 1: asymmetrization-kinetic resolutions. *Tetrahedron: Asymmetry* 8: 2812–2817

74 Gaziola L, Bornscheuer U, Schmid RD (1996) A rapid and effective separation of enantiomers of glycerol derivatives by gas chromatography and their lipase-catalyzed biotransformation. *Enantiomer* 1: 49–54

75 Lampe TFJ, Hoffmann HMR, Bornscheuer UT (1996) Lipase mediated desymmetrization of *meso* 2,6-di(acetoxymethyl)tetrahydropyran-4-one derivatives - an innovative route to enantiopure 2,4,6-trifunctionalized *C*-glycosides. *Tetrahedron: Asymmetry* 7: 2889–2900

76 Bornscheuer U, Capewell A, Wendel V, Scheper T (1996) On-line determination of the conversion in a lipase-catalyzed kinetic resolution in supercritical carbon dioxide. *J Biotechnol* 46: 139–143

77 Bornscheuer U, Herar A, Kreye L, Wendel V, Capewell A, Meyer HH, Scheper T,

Kolisis FN (1993) Factors affecting the lipase catalyzed transesterification reactions of 3-hydroxy esters in organic solvents. *Tetrahedron: Asymmetry* 4: 1007–1016

78 Weissfloch ANE, Kazlauskas RJ (1995) Enantiopreference of lipase from *Pseudomonas cepacia* toward primary alcohols. *J Org Chem* 60: 6959–6969

79 Heidt M, Bornscheuer U, Schmid RD (1996) Studies on the enantioselectivity in the lipase-catalyzed synthesis of monoacylglycerols from the isopropylidene glycerol. *Biotechnol Tech* 10: 25–30

80 Lampe TFJ, Hoffmann HMR (1996) Asymmetric synthesis of the C(10)-C(16) segment of the bryostatins. *Tetrahedron Lett* 37: 7695–7698

81 Kazlauskas RJ, Weissfloch ANE, Rappaport AT, Cuccia LA (1991) A rule to predict which enantiomer of a secondary alcohol reacts faster in reactions catalyzed by cholesterol esterase, lipase from *Pseudomonas cepacia*, and lipase from *Candida rugosa*. *J Org Chem* 56: 2656–2665

82 Wünsche K, Schwaneberg U, Bornscheuer UT, Meyer HH (1996) Chemoenzymatic route to β-blockers via 3-hydroxy esters. *Tetrahedron: Asymmetry* 7: 2017–2022

83 Chen CS, Wu SH, Girdaukas G, Sih CJ (1987) Quantitative analyses of biochemical kinetic resolution of enantiomers. 2. Enzyme-catalyzed esterifications in water-organic solvent biphasic systems. *J Am Chem Soc* 109: 2812–2817

84 Anthonsen HW, Hoff BH, Anthonsen T (1995) A simple method for calculating enantiomer ratio and equilibrium constants in biocatalytic resolutions. *Tetrahedron: Asymmetry* 6: 3015–3022

85 Rakels JLL, Caillat P, Straathof AJJ, Heijnen JJ (1994) Modification of the enzyme enantioselectivity by product inhibition. *Biotechnol Prog* 10: 403–409

86 van Tol JBA, Jongejan JA, Duine JA (1995) Description of hydrolase-enantioselectivity must be based on the actual kinetic mechanism: analysis of the kinetic resolution of glycidyl (2,3-epoxy-1-propyl) butyrate by pig pancreas lipase. *Biocatal Biotransform* 12: 99–117

87 van Tol JBA, Kraayveld DE, Jongejan JA, Duine JA (1995) The catalytic performance of pig pancreas lipase in enantioselective transesterification in organic solvents. *Biocatal Biotransform* 12: 119–136

7 Peptide Synthesis in Non-Aqueous Media

Pere Clapés, Gloria Caminal, Josep A. Feliu and Josep López-Santín

Contents

Methods and Tools in Biosciences and Medicine
Methods in non-aqueous enzymology, ed. by M. N. Gupta
© 2000 Birkhäuser Verlag Basel/Switzerland

1   Introduction

The interest in the application of proteases for peptide synthesis has been growing since 1979, when Morihara et al. converted porcine insulin to human insulin by a two-step enzymatic process, which has been employed by Novo Nordisk A/s (Denmark) to prepare semisynthetic human insulin [1]. Another key point in the development of the enzymatic approach has been the successful synthesis of the artificial sweetener aspartame [2]. Nowadays, the synthesis of peptides catalyzed by proteases is a complementary tool to the well-established chemical procedures. The potential advantages of an enzymatic synthesis are: mild operation conditions, minimal solvent consumption, reduction of undesired by-products, minimal protection and, in many cases, easy scaling-up [3, 4]. The main drawbacks are the lack of systematic rules for general applications, possible deactivation of the biocatalyst, as well as the few applications still to date in the synthesis of large peptides.

An outline of the different steps for a rational approach to a desired enzymatic peptide synthesis can be drawn as follows:
- Selection of the suitable protease(s) as biocatalyst
- Selection of a thermodynamically or kinetically controlled approach
- Preparation of starting amino acid or peptide derivatives
- Definition of the reaction medium as well as the compatible biocatalyst preparation (reaction medium engineering)
- Study of the reaction kinetics, yields and factors influencing them such as enzyme loading, substrate concentration, solvent and water concentration or water activity
- Scale-up to the desired preparative level

The above steps must be accomplished for each individual peptide bond to be synthesized. For stepwise or convergent strategies, a previous selection of the appropriate synthetic scheme has to be devised. When the conditions for each peptide synthesis have been determined, an appropriate procedure has to be proposed by minimizing the number of steps and considering that the products of one synthesis will be the substrates for the next step. Some of the indicated aspects are discussed below.

Proteases, in addition to their regio and stereoselectivity, usually exhibit a preference for distinct amino acid side chains placed nearby the scissile peptide bond or even for residues located further from this position. Thus, this preference has to be taken into account first for a selection of a suitable protease. The enzymes more often used are those commercially available like α-chymotrypsin, trypsin, papain, thermolysin or carboxypeptidase Y, among others, which are obtained in crystalline form or in partially purified preparations.

The enzyme-catalyzed synthesis of peptide bonds can be carried out under thermodynamic or kinetic control [5]. In the former case, a direct reversion of the protein hydrolysis occurs, according to the scheme (Fig. 1).

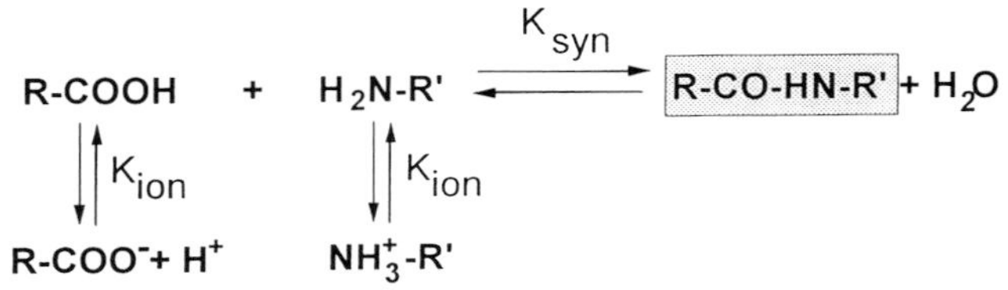

$$R\text{-COOH} \quad + \quad H_2N\text{-}R' \quad \underset{\longleftarrow}{\overset{K_{syn}}{\longrightarrow}} \quad \boxed{R\text{-CO-HN-}R'} + H_2O$$

$$\big\updownarrow K_{ion} \qquad\qquad \big\updownarrow K_{ion}$$

$$R\text{-COO}^- + H^+ \qquad NH_3^+\text{-}R'$$

Figure 1 Thermodynamically controlled peptide bond formation.

It has to be noticed that only non-ionized substrates (amino acids or peptides) participate in the synthesis equilibrium. There are two ways to shift equilibrium towards the desired peptide coupling: i) a modification of K_{ion} can be accomplished by reducing the ionization of the α-carboxylic and α-amino groups. This can be done by using a less-polar cosolvent in water as reaction medium; and ii) application of the principle-of-mass action by reducing either the concentration of the peptide product (i.e. by precipitation or selective extraction in a biphasic system) or the concentration of water (i.e. organic media).

The synthesis under kinetic control is based on the fact that serine and cysteine proteases are able to form a covalent acyl-enzyme complex with an amino acid or peptide carboxyl ester as acyl donor. The acyl-enzyme complex then undergoes a competitive deacylation by water (to yield the acid) or by an added nucleophile to give the desired product (Figure 2). The obtained peptide yield is dependent on both the concentration and reactivity of the nucleophile as well as the degree of hydrolysis of the peptide product. Hence, it is important to work at low water concentration to favor the synthesis of the peptide bond and minimize the proteolytic activity. The advantage is that kinetically-controlled syntheses have faster reaction rates than that thermodynamically controlled ones and less enzyme is needed.

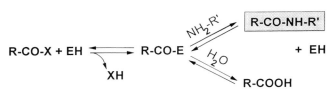

Figure 2 Kinetically controlled peptide synthesis by serine and cysteine proteases (EH) with R-COO-X as the acyl-donor and H_2N-R' as the nucleophile (R and R' are protected amino acid or peptide derivatives).

In summary, reaction medium engineering becomes one of the keys for obtaining good synthetic yields of the peptide product using either thermodynamic or kinetic control approaches [6–8]. The reaction systems can be classified as follows:

- Cosolvent systems: Water soluble solvents such as alcohols or dimethylformamide (DMF) can be employed. Nevertheless the activity of free or immobilized proteases in this medium will be affected by the presence of the organic solvent.
- Biphasic systems: This approach can be used when both acyl-donor and nucleophile are soluble in water and the product is mainly soluble in the organic phase. The reaction takes place in the aqueous phase and the enzyme can be used either free or immobilized by covalent attachment. Mass transfer limitations and deactivation of proteases at the interphase are the major drawbacks of this reaction medium. This problem can be overcome by using appropriate surfactants to reduce the interfacial tension [9].
- Low-water content systems: Proteases are active in reaction media with very low water concentration. They can be employed either as enzyme powder or deposited onto a support [10, 11]. The thermodynamic water activity (a_w) in these systems is directly related with the enzyme activity and it also influences the ratio synthesis rate/hydrolysis rate in kinetically controlled synthesis. Low-water content systems such as organic media have a wide range of applications in oligopeptide synthesis, because they are normally compatible with the solubility requirements of amino acid and peptide derivatives [12].

As pointed out before, most of the published applications on enzymatic peptide bond formation are related with the synthesis of short peptides. There are some good reviews [3, 4, 13], emphasizing the potential applications of the technique in: i) syntheses of bioactive peptides and analogs; ii) enzymatic modifications of biosynthetic precursors; and iii) synthesis of semisynthetic proteins. One of the applications to be mentioned is the use of proteases as catalysts in fragment condensation. The enzymatic condensation of minimally protected peptide segments to produce long polypeptides and small proteins can overcome some of the current limitations of conventional chemical solution and solid phase methods [14].

In this chapter, the methodology for the enzymatic synthesis of the following selected biologically active peptides in nonaqueous systems is presented: Endogenous opiod peptide Leu-enkephalin derivatives (Z-Tyr-Gly-Gly-Phe-Leu-R_1, where R_1: NH$_2$, OEt) [15], cholecystokinin COOH-terminal pentapeptide CCK-8 (5–8) (H-Gly-Trp-Met-Asp-Phe-NH$_2$) with intestinal and brain functions [16] and carboxyglutamic acid containing peptides (Z-γ,γ'-di-*tert*-butyl-L-Gla-L-XX-OR$_1$) related with osteocalcins [17], proteins of the coagulation cascade [18] and conantokins [19]. Amino acid or peptide modifications to get active acyl-donors or nucleophiles, as well as the protecting groups employed are also discussed.

2   Materials

Chemicals
- Amino acid derivatives were purchased from SIGMA (St. Louis, MO, USA), Bachem (Bubendorf, Switzerland) and NOVA Biochem (Lufelfingen, Switzerland) except those prepared in our laboratory (see Protocols)
- Acetonitrile and ethyl acetate were of high performance liquid chromatography (HPLC) grade
- Milli-Q Water was used in all the studies

All other reagents used were of analytical grade.

Enzymes
- α-chymoptrypsin (EC 3.4.21.1) from bovine pancreas (crystallized, lyophilized powder 5 U/mg N-α-acetyl-L-tyrosine ethyl ester (ATEE) assay). Merck (Barcelona, Spain)
- Papain (EC 3.4.22.2) from Carica papaya (2x crystallized powder, 10–20 U/mg N-α-benzoyl-L-arginine ethyl ester (BAEE) assay). Fluka (Buchs, Switzerland)
- Thermolysin (EC 3.4.24.2) from *Bacillus thermoproteolyticus rokko* (crystallized and lyophilized powder containing calcium and sodium buffer salts, 50–100 U/mg protein, casein assay). Sigma Chemical Co. (St. Louis, MO, USA)
- Bromelain (EC 3.4.22.4) (crude powder containing approximately 50% of protein, 2–4 U/mg of protein, N-α-carbobenzoxy-L-Lysine p-nitrophenyl ester assay). Sigma Chemical Co. (St. Louis, MO, USA).

Equipment
- Support: Celite R-633 particle size 297–595 μm, mean pore diameter 6500 nm, specific surface area (BET method) 2.25 m^2/g. Manville (Pleasanton CA, USA)
- Support: Celite-545 particle size 20–45 μm, mean pore diameter 17000 nm, specific surface area (BET method) 2.19 m^2/g. Fluka (Buchs, Switzerland)
- Support: Polyamide 6 particle size 200–1000 μm, mean pore diameter 1860 nm, specific surface area (BET method) 47 m^2/g. Generous gift from Akzo (Obernburg, Germany)
- Preparative HPLC runs were performed on a Waters (Milford, MA, USA) Prep LC 4000 pumping system and a Waters PrePack® 1000 module fitted with a PrePack® (Waters) column (47 x 300 mm) filled with either VYDAC™ C_{18}, 300 Å, 15–20 μm stationary phase or Bondapack C_{18}, 300 Å, 15–20 μm stationary phase
- The analytical HPLC systems were: Merck-Hitachi Lichrograph system (Darmstadt, Germany) fitted with a Lichrocart 250–4 HPLC cartridge, 250 x 4 mm, filled with Lichrospher® 100, RP-18, 5 μm (Merck) and Kontron Analytical System fitted with a Vydac (The Separation Group, Hesperia, CA) C_{18} column (4.6 x 250 mm, 5μm).

- Measurements of water concentration were performed in a Karl-Fisher moisture analyzer MKC-210 Coulometer from Kyoto Electronics, equipped with a moisture evaporator model ADP-351 for solid samples
- Product lyophilization as well as the enzyme-support drying was performed in a freeze-drying equipment with a CHRIST Alfa 2–4 (Braun, Osterode, Germany) condenser connected to an Edwards 28 two-stage vacuum pump (Sussex, UK)
- Reactions were stirred at constant temperature in a Certomat WR thermostated reciprocal shaking-baths (Braun, Osterode, Germany)

Solutions
- Tris(hydroxymethyl)aminomethane-HCl buffer (Tris-HCl), 50 mM at pH 9.0
- Boric acid-borate buffer, 0.1 M at pH 8.2 and pH 9.0
- 3-Morpholinopropanesulfonic acid (MOPS) buffer, 50 mM at pH 7.0

3   Methods

In this section, general procedures such as enzyme deposition, reaction medium engineering and mode of operation (synthesis, analysis and purification) for enzymatic peptide synthesis in organic media are presented.

3.1   Immobilization

Deposition of enzymes onto solid supports
The optimal enzyme loading (mg of enzyme/g of support), the one that gives the maximal initial reaction rate, depends on both the support and enzyme selected. As a first trial, prior to optimization studies, 30 mg of enzyme per g of support may be used.

The protease (30 mg) is dissolved in the appropriate buffer (1 mL) at its optimal pH, mixed thoroughly with the support (1 g) and dried. Deposition of papain and bromelain requires the presence of a reducing agent (1,4 dithio-DL-threitol; 30 mg) to prevent oxidation of the thiol group in the active site. Both enzymes are also inhibited by heavy metal ions, thus ethylenediaminetetraacetic acid (EDTA, 0.5 mM) has to be added to the buffer solution.

Once the enzyme solution and the support have been mixed thoroughly, the mixture must be dried. The most commonly used method for drying the enzyme-support mixture is evaporation under vacuum. Alternatives could be lyophilization or direct equilibration at a fixed water activity (see 3.2).

3.2 Fixing initial water activity

When reaction performances in different reaction systems are going to be compared, it becomes necessary to work at a fixed water activity (a_w). Sometimes, it is difficult to monitor the a_w and keep it constant during the reaction. Water activity could vary when water participates in the reaction. It should be pointed out that in carrying the reactions under kinetic control no production of water occurs, unlike in the case of thermodynamic control, where water is a reaction product. On the other hand, during a hydrolysis reaction water is a reactant and it is consumed along with the substrate. If initial reaction rates are to be evaluated, then a constant value of water activity at the beginning of the reaction can be fixed easily.

One of the methods to fix initial water activity is to pre-equilibrate the reaction system, solvent, reactants and enzyme, with an atmosphere of fixed a_w value generated by a convenient saturated salt solution [20].

To assure an initial fixed a_w, the following steps are necessary:

Catalyst equilibration
The solid catalyst (support plus enzyme, 1 g) is placed into an open vessel that is kept into a closed flask (see Figure 3) containing a saturated salt solution that generates an atmosphere of fixed a_w. The equilibrium is attained more quickly if a high surface/volume ratio of the solid is chosen. The system must be placed at a constant temperature that has to be the same as the reaction temperature used, normally 25 °C. Equilibrium is achieved when the measured water content of the solid remains constant. Equilibration time depends on many factors, among them: support, vessel size, previous water content, desired a_w. For instance, the equilibration of 1g of α-chymotrypsin-polyamide at a_w=0.113 (LiCl) took 4 days using a vessel with a surface/volume ratio of 0.5.

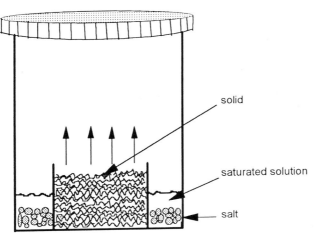

Figure 3 Experimental set-up to equilibrate the deposited biocatalyst at a fixed a_w.

solid

saturated solution

salt

Reaction media equilibration

Substrates are dissolved into the selected solvent, and the open reactor is placed into the well-closed flask with the adequate saturated salt solution. The equilibration of 2 mL of ethyl acetate containing the substrates at $a_w=0.113$ (LiCl) took 24 hrs approximately.

This method is impractical for other than small-scale reactors and for water miscible solvents. An alternative procedure is suggested. When the relationship between a_w and water concentration is known, i.e. experimentally determined or estimated by a thermodynamic model, water may be added to anhydrous solvent to reach the desired water activity. In such cases, the catalyst is equilibrated as indicated above and the a_w of the reaction medium adjusted by adding the corresponding amount of water (or buffer).

3.3   Substrate preparation

Amino acid and peptides Carboxamidomethyl ester (Cam) derivatives

Acyl donor ester substrates were either directly purchased or synthesized. Usually, all the N-α protected amino acid or peptides Cam derivatives were prepared in our lab by the cesium salt method [21] (see Protocol 3, Z-Gly-Trp-OCam).

Removal of N-α protecting groups

In convergent strategies the N-α protecting group of the C-terminal fragments has to be removed. The N-α protecting groups used in the examples described were benzyloxycarbonyl (Z) and *tert*-butyloxycarbonyl (Boc). The methodology for their elimination is well documented in the literature [22, 23]. Benzyloxycarbonyl group was removed by catalytic hydrogenation using palladium-over-charcoal under hydrogen atmosphere or by acidolysis in the presence of liquid hydrogen fluoride. *Tert*-butyloxycarbonyl and *tert*-butyl (OBu^t) groups were removed by acidolysis using trifluoroacetic acid.

3.4   Synthesis reactions

Reactions were carried out in closed Erlenmeyer flasks (100–250 mL) with reciprocal shaking or in closed reactors (50–500 mL), equipped with mechanical stirring, both at a constant temperature of 25 °C.

Enzymatic reactions were started by adding the corresponding protease deposited onto the support to the organic medium containing the substrates and a controlled amount of aqueous buffer or at a fixed initial thermodynamic activity of water if necessary.

The amount of reactants and products at different reaction times was determined by HPLC. Sampling was performed as follows. The shaking or stirring was stopped and the solid catalyst was allowed to deposit at the bottom of the reactor. Samples (50–100 µL) were withdrawn from the supernatant, mixed with acetic acid (20 µL) to stop any further enzymatic reaction, dissolved in HPLC eluent (250–1000 µL), depending on the extinction coefficient of the compounds, and analyzed subsequently.

3.5 Separation and purification

Separation

Once the limiting reactant was completely consumed or the maximum product yield was detected, the reaction was stopped by filtering off the immobilized preparation. When the product precipitated during the reaction, it was dissolved with methanol or ethanol prior to being filtered off. The filtrate was worked up as follows: The solvent was evaporated under vacuum. The residue was dissolved in ethyl acetate or dichloromethane and washed successively with aqueous solutions of citric acid 5% (w/v), $NaHCO_3$ 10% (3x) and saturated NaCl solution (1x). Normally the ratio organic phase/aqueous phase was 1. Then, the organic layer was dried over anhydrous sodium sulfate and the solvent evaporated under reduced pressure. If the purity of the product is enough, which usually happened with the intermediates, then the product was ready to be used in the next synthetic step. If not, further purification by crystallization, flash chromatography on silica or preparative HPLC was performed.

Purification
- Crystallization: Crystallization was usually performed using ethyl acetate/ hexane or ethanol/hexane mixtures. The relative proportion of the solvents depends on the product considered.
- Flash chromatography on silica: When the product was soluble enough in ethyl acetate or ethyl acetate/hexane mixtures, flash chromatography on silica was the technique selected to further purify the product when required. The procedure is well-described in the literature and it was followed without any significant modification [24].
- Preparative HPLC: When the products were soluble in polar solvents such as methanol, acetonitrile or mixtures of these solvents with water, then preparative HPLC was the method of choice. This technique can also be applied without the initial work up of the filtrate.

Purification by preparative HPLC was performed as follows: The crude peptides were loaded onto the preparative column (see 2.1). Products were eluted using CH_3CN gradients (0.2% CH_3CN/min) in triethylamine-phosphoric acid buffers (TEAP) at different pH values or in aqueous 0.1% trifluoroacetic acid (TFA). The flow rate was 100 mL/min and the products were detected either at 215 nm or 225 nm. Analysis of the fractions was accomplished under isocratic conditions, using VYDAC C-18, 5µ, 0.46 x 25 cm column or Lichrosphere RP-18 5µ 0.46 x 25 cm column, eluted with different proportions of (A) 0.1% aqueous

TFA and (B) 0.085% TFA in water: CH_3CN 1:4, flow rate 1.5 mL/min and detection at 215 nm. Pure fractions were pooled and desalted on the same preparative cartridge by a CH_3CN gradient in 0.1% (v/v) TFA. The eluates were then lyophilized

3.6 HPLC analysis

All the reactions were followed by HPLC analysis. The solvent system used was the following: solvent (A) aqueous 0.1% TFA/(B) 0.085% TFA in CH_3CN or 0.085% TFA in $CH_3CN:H_2O$ 4:1; flow rate 1 mL/min; detection 254 nm or 215 nm Products and substrates were eluted using gradients of solvent B or under isocratic conditions.

3.7 Protocols

In this section, the use of nonaqueous media for protease-catalyzed synthesis of peptides is described for the following examples: Leu-enkephalin derivatives (Z-Tyr-Gly-Gly-Phe-Leu-R_1 where R_1: NH_2, OEt), CCK-8 (5–8) pentagastrin (H-Gly-Trp-Met-Asp-Phe-NH_2) and the diastereoselective synthesis of carboxyglutamic acid containing dipeptide derivatives (Z-γ,γ'-di-*tert*-butyl-L-Gla-L-XX-OR_1). The synthesis of Leu-enkephalin derivatives and cholecystokinin C-terminal pentapeptide (CCK-8 (5–8)) was carried out at preparative level and the synthetic schemes used are presented in Figures 4, 5 and 6. The diastereoselective synthesis of carboxyglutamic acid containing dipeptides was carried out at analytical level except in the case of Z-γ,γ'-di-*tert*-butyl-L-Gla-L-Leu-OEt which was produced at gram scale.

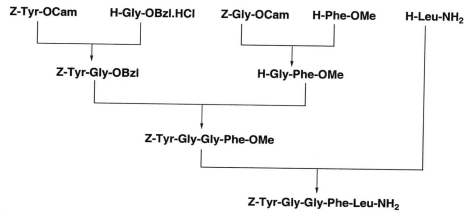

Figure 4   Synthetic scheme of the enzymatic synthesis of Z-[Leu]-enkephalin amide (Z-Tyr-Gly-Gly-Phe-Leu-NH_2) derivative in organic media.

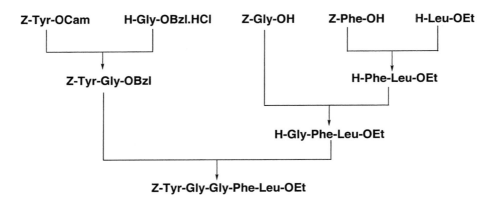

Figure 5   Synthetic scheme of the enzymatic synthesis of Z-[Leu]-enkephalin ethyl ester (Z-Tyr-Gly-Gly-Phe-Leu-OEt) derivative in organic media.

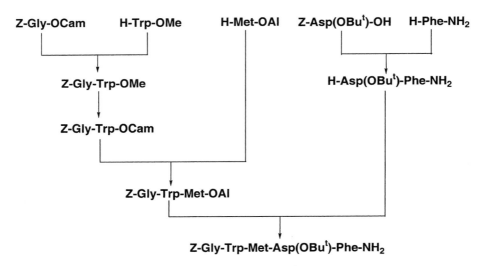

Figure 6   Synthetic scheme of the enzymatic synthesis of CCK-5 (4–8) (Z-Gly-Trp-Met-Asp(O-Bu^t)-Phe-NH$_2$) derivative in organic media.

Protocol 1   Enzymatic synthesis of Z-Tyr-Gly-Gly-Phe-Leu-NH$_2$

1. *Z-Tyr-Gly-OBzl*: Z-Tyr-OCam (2.9 g, 8.0 mmol) and H-Gly-OBzl.HCl (2.7 g, 13.6 mmol) were dissolved in acetonitrile (100 mL) containing Tris-HCl 50 mM, pH 7.8 buffer (1% v/v) and triethylamine (2.5 mL 17.7 mmol). Then a preparation of α-chymotrypsin on Celite (6 g, 20 mg of α-chymotrypsin/g of celite) was added. After 12 hrs the HPLC dipeptide yield was 92%. The product was worked up (see 3.5) and crystallized (hexane/ethyl acetate) obtaining 2.9 g (79%) of the dipeptide derivative. HPLC reaction monitoring: 40% B for 5 min and gradient from 40 to 80% B over 20 min.

2. *Z-Gly-Phe-OMe:* Z-Gly-OCam (2.13 g, 8 mmol) and H-Phe-OMe (2.87 g, 16 mmol) were dissolved in separate flasks with ethyl acetate (50 mL) containing mercaptoethanol (0.15 mL). These solutions were equilibrated with a saturated salt solution of KNO_3 for two days (a_w=0.936). Then, the solutions containing the substrates were mixed and papain on celite (1 g, 15 mg of papain/g of Celite) previously equilibrated at the same water activity was added. After 2 hrs of reaction time the yield by HPLC was 98%. The reaction was worked up (see 3.5) and the pure dipeptide was obtained as a pale yellow oil (2.7 g, 91%). HPLC reaction monitoring: 30% B for 5 min and gradient from 30 to 70% B over 15 min.

3. *Z-Tyr-Gly-Gly-Phe-OMe:* H-Gly-Phe-OMe.TFA (1.7 g, 4.9 mmol) and Z-Tyr-Gly-OBzl (1.5 g, 3.2 mmol) were dissolved in acetonitrile (100 mL) containing boric acid-borate 0.1 M, pH 8.2 buffer (4% v/v) and triethylamine (0.75 mL). To this solution, a preparation of papain on Celite (1.5 g, 15 mg papain/g Celite) was added. After 6 hrs of reaction time, the yield by HPLC was 80%. After the work-up (see 3.5), the pure tetrapeptide was obtained by crystallization (ethyl acetate/hexane), yielding a white solid (1.4 g, 74%). HPLC reaction monitoring: gradient elution from 30 to 80% B over 50 min.

4. *Z-Tyr-Gly-Gly-Phe-Leu-NH$_2$:* Z-Tyr-Gly-Gly-Phe-OMe (0.56 g, 0.95 mmol) and H-Leu-NH$_2$ (0.25 g, 1.91 mmol) were dissolved in acetonitrile (50 mL) containing Tris-HCl 50 mM, pH 7.8 buffer (4% v/v). To this solution, a preparation of α-chymotrypsin on Celite (2 g, 15 mg α-chymotrypsin/g of Celite) was added. After 1.5 hrs the product yield by HPLC was 85% and no Z-Tyr-Gly-Gly-Phe-OMe was detected. HPLC reaction monitoring: 40% B for 5 min and gradient from 40 to 80% B over 20 min.

5. The crude peptide was purified by preparative HPLC as indicated above (see 3.5). First the peptide was chromatographed using a TEAP system at pH 6.7 and gradient elution from 21 to 45% CH_3CN over 60 min. Analysis of the fractions was performed as described in methods using 45% B for elution. Pure fractions were pooled, loaded onto the cartridge and eluted using 0.1% TFA system under gradient conditions 27 to 43% CH_3CN in 30 min. Impure fractions from the previous TEAP run were re-chromatographed on the TFA system using a gradient from 31 to 39% CH_3CN in 25 min. The best fractions from both 0.1% TFA runs were pooled and lyophilized (0.37g, 66%). Purity 99.7% by HPLC.

Protocol 2   Enzymatic synthesis of Z-Tyr-Gly-Gly-Phe-Leu-OEt

1. *Z-Phe-Leu-OEt:* Z-Phe-OH (2.1 g, 7.1 mmol) and H-Leu-OEt (3.4 g, 21.3 mmol) were dissolved in acetonitrile (100 mL) containing MOPS (Na$^+$) 50 mM, pH 7 buffer (4% v/v), and incubated in the presence of a preparation of thermolysin on Celite (2 g, 30 mg thermolysin/g of Celite). After 6 hrs, the dipeptide yield by HPLC was 93%. The reaction was worked up as described (see 3.5). Crystallization (ethyl acetate/hexane) yielded a white solid chromatographically homogeneous product (2.6 g, 83%). HPLC reaction monitoring: 50% B for 5 min and gradient from 50 to 85% B over 5 min.

2. *Z-Gly-Phe-Leu-OEt:* Z-Gly-OH (2.5 g, 12.0 mmol) and H-Phe-Leu-OEt (2.5 g, 8.0 mmol) were dissolved in acetonitrile (100 mL) containing MOPS (Na$^+$) 50 mM, pH 7 buffer (4% v/v). To this solution, a preparation of thermolysin on Celite (4 g, 30 mg thermolysin/g of Celite) was added. After 24 hrs the tripeptide yield by HPLC was 90%. The product was worked up as described (see 3.5). Crystallization (ethyl acetate/hexane) rendered the pure tripeptide (3g, 76%). HPLC reaction monitoring: 45% B for 5 min and gradient from 45 to 80% B over 10 min.

3. *Z-Tyr-Gly-Gly-Phe-Leu-OEt:* Z-Tyr-Gly-OBzl (0.40 g, 0.9 mmol) (Protocol 1, Z-Tyr-Gly-OBzl) and H-Gly-Phe-Leu-OEt (0.25 g, 0.7 mmol) were dissolved in acetonitrile (30 mL) containing boric acid-borate 0.1 M, pH 9 buffer (4% v/v), and mercaptoethanol (0.25 mL). To this solution, a preparation of bromelain on Celite (3 g, 80 mg bromelain/g of Celite) was added. After 24 hrs the pentapeptide yield by HPLC was 97%. HPLC reaction monitoring: 50% B for 5 min and gradient from 50 to 74% B over 11 min. The product was worked up as described above and further purified by flash cromatography on silica (see 3.5). First, the crude was eluted with ethyl acetate to remove the excess of Z-Tyr-Gly-OBzl. Then, the pure pentapeptide was eluted with ethyl acetate/methanol 19:1 (v/v) yielding 0.35 g (70%) of pentapeptide. Purity 99.9 by HPLC.

Protocol 3   Enzymatic synthesis of H-Gly-Trp-Met-Asp-Phe-NH$_2$

1. *Z-Gly-Trp-OCam:* Z-Gly-OCam (2.6 g, 98 mmol) and H-Trp-OMe (3.2 g, 147 mmol) were dissolved in acetonitrile (96 mL) containing boric acid-borate 0.1 M, pH 8.2 buffer (3.7 mL) and β-mercaptoethanol (0.3 mL). To this solution a papain-Celite preparation (2 g, 15 mg papain/g of Celite) was added. After 1hr of reaction time the product yield measured by HPLC was 98%. The product was worked up as described above and the residue was lyophilized yielding a white solid (3.7 g, 93%). HPLC reaction monitoring: Solvent system (A): aqueous 0.1% (v/v) orthophosphoric acid, (B): buffer A/CH$_3$CN (3:17 v/v); isocratic elution 75% B, flow rate 1.0 mL/min, detection 215 nm. The synthesis of Z-Gly-Trp-OCam was accomplished by means of the cesium salt method starting from Z-Gly-Trp-OH. Z-Gly-Trp-OH was obtained by α-chymotrypsin-catalyzed hydrolysis of Z-Gly-Trp-OMe. Z-Gly-Trp-OMe (7.0 g, 17.1 mmol) was dissolved in acetonitrile (133 mL) containing Tris-HCl 50 mM, pH 9 buffer (16 mL) and triethylamine (1mL). To this solution a pre-

paration of α-chymotrypsin on Celite (2 g, 30 mg α-chymotrypsin/g Celite) was added. The reaction was completed in 24 hrs and no ester substrate was detected by HPLC. The residue was freeze-dried yielding a white solid (6.39 g, 94%). Z-Gly-Trp-OH (3.63 g, 9.2 mmols) was dissolved in ethanol-water 70:30 (50 mL) and cooled down till 0 °C. Then, a solution of cesium carbonate (1.62 g, 4.6 mmols) in water (10 mL) was slowly added. After the solution was stirred for 5 min, the solvent was evaporated to dryness in vacuum, and the residue lyophilized. The white solid obtained was suspended in dimethylformamide (DMF) and α-chloroacetamide (1.29 g, 13.8 mmols) was added. After the mixture was stirred for 96 hrs, the reaction was worked up as described (see 3.5). Crystallization (ethyl acetate/hexane) rendered 6.89 g (93%) of pure product.

2. *Z-Asp(OBu^t)-Phe-NH$_2$*: Z-Asp(OBu^t)-OH (3.8 g, 11.8 mmol) and H-Phe-NH$_2$ (2.9 g, 17.8 mmol) were dissolved in acetonitrile (96 mL) containing MOPS 50 mM pH 7.0 buffer (4 mL). To this solution, a preparation of thermolysin on Celite-R633 (2 g, 30 mg thermolysin/g of Celite) was added. The reaction was completed in 6 hrs and the product yield by HPLC was 99.8%. The product precipitated in the reaction medium. The reaction mixture was worked up as described (see 3.5) yielding a white solid (5.12 g, 92%). HPLC reaction monitoring: gradient elution from 30 to 90% B in 25 min.

3. *Z-Gly-Trp-Met-OAl*: Z-Gly-Trp-OCam (3.4 g, 7.5 mmol) and H-Met-OAl (6.2 g, 33 mmol) were dissolved in acetonitrile (100 mL) containing Tris-HCl 50 mM, pH 9 buffer (0.5 mL) and triethylamine (0.5 mL). To this solution, α-chymotrypsin on Celite preparation (2 g, 8 mg α-chymotrypsin/g of Celite) was added. After 22 hrs, (67% yield by HPLC) the reaction was worked up as described (see 3.5). The product was further purified by flash chromatography on silica (eluent: hexane:ethyl acetate 1:4) yielding 2.2 g (52%) of pure product. HPLC reaction monitoring: gradient elution from 40% B to 80% B over 24 min, detection 278 nm.

4. *H-Gly-Trp-Met-Asp-Phe-NH$_2$*: Z-Gly-Trp-Met-OAl (3.26 g, 6 mmol) and H-Asp(OBu^t)-Phe-NH$_2$ (2.9 g, 8.6 mmol) were dissolved in acetonitrile (189 mL) containing Tris-HCl 50 mM, pH 9 buffer (0.95 mL) and triethylamine (0.95 mL). To this solution, a preparation of α-chymotrypsin on polyamide (6.4 g, 30 mg α-chymotrypsin/g polyamide) was added. Previously, the enzyme-support preparation was pre-equilibrated at a_w=0.113 (see 3.2). After 4 days, the reaction was stopped and the product yield by HPLC was 76%. In this case the product precipitated during the reaction. HPLC reaction monitoring: gradient elution 30 to 80% B in 30 min.

5. Deprotection of the β-tert-butyl protecting group of the aspartic acid residue and the de-acylation of Z-pentapeptide was performed on the crude peptide.

6. Product purification by HPLC was carried out as described in 3.5. First the peptide was chromatographed using a TEAP system at pH 2.25 and gradient elution from 14 to 26% CH$_3$CN over 60 min. Analysis of the fractions was performed as described using 39% B. Pure fractions were pooled, desalted and lyophilized yielding (1.2 g, 39%) of pure (99.8 by HPLC) pentapeptide.

Protocol 4 Diastereoselective enzymatic synthesis of Z-γ,γ'-di-*tert*-butyl-L-Gla-L-AA-OR$_1$ (R$_1$: NH$_2$, OEt)

1. Reactions at test level were carried out as follows: Z-γ,γ'-di-*tert*-butyl-D,L-Gla-OMe (34.8 mg, 0.08 mmol) and the nucleophile (H-AA-OR) (0.06 mmol, 50% excess respect to the L isomer of the donor ester) were dissolved in oxygen-free acetonitrile or ethyl acetate (2 ml) containing 0.1 M boric acid-borate pH 8.5 buffer (2% v/v), β-mercaptoethanol (1% v/v) and triethylamine when necessary (0.07 mmol). To this solution papain deposited on Celite (200 mg) was then added and the mixture kept under argon atmosphere. The yields obtained for different amino acid nucleophiles are given in Table 1.

Table 1 *Papain-catalyzed diastereoselective synthesis of carboxyglutamic acid (Gla) containing peptides.*

Dipeptide yields and reaction time obtained using different amino acid derivatives as nucleophiles. Reaction conditions are described in Protocol 4.

Reaction	Nucleophile (mmol)	TEA (mmol)	Dipeptide Yield (%)	Reaction Time (hr)
1	H-L-Ala-NH$_2$	—	57	24
2	H-L-Ile-NH$_2$	0.07	89	48
3	H-L-Val-NH$_2$	0.07	81	48
4	H-L-Phe-NH$_2$	–	89	24
5	H-Gly-NH$_2$	0.07	73	48
6	H-L-Leu-NH$_2$	–	96	24
7	H-L-Tyr-OEt	–	89	24

2. Z-γ,γ'-di-tert-butyl-L-Gla-L-Leu-OEt: Z-γ,γ'-di-*tert*-butyl-D,L-Gla-OMe (2.25 g, 5 mmol) and H-Leu-OEt (0.73 g, 3.7 mmol) were dissolved in acetonitrile (100 ml) containing 0.1 M boric acid-borate pH 8.2 buffer (0.5% v/v) and triethylamine (0.5 ml, 3.7 mmol). This solution was degassed with N$_2$ for about 3 min and kept under argon atmosphere. To this solution a papain-celite preparation (10 g, 100 mg of crude papain/g of celite-545) was added. The mixture was placed on a reciprocal shaker (130 rpm) at 25°C. After 24 hrs all the Z-γ,γ'-di-*tert*-butyl-L-Gla-OMe was consumed and the yield by HPLC was 99% with respect to L isomer. The reaction mixture was worked up as described and the residue was further purified by preparative HPLC on a Delta-Pack C-4, 15 μm (5 x 30 cm) cartridge. The crude peptide was chromatographed using a CH$_3$CN gradient in aqueous TFA (52 to 68% CH$_3$CN over 30 min). Analysis of the fractions was performed as described (see 3.5) using 64% B. Pure fractions were pooled and freeze dried to give an oil (0.89 g, 60%) with a purity of 97% by HPLC.

4   Troubleshooting

- *One of the problems when dealing with oligopeptides is their solubility in the reaction medium.* Examples are described in the literature in which substrates are used mainly in solid form providing product yields ranging from 90 to 99% [25, 26]. This is true when simple amino acid derivatives are used, whereas for peptide fragments this approach was useless in our hands.

- *Working with polyamide, substrate and product adsorption or deposition onto the support can be observed during the reaction.* Hence, their time evolution cannot be followed by direct HPLC analysis of the liquid mixture. The products must be extracted previously. In test experiments a mixture of acetic acid:methanol:dimethylsulfoxide (1:4:5) was suitable. In preparative reactions an extraction with acetic acid:methanol 1:4 is recommended.

- *Using thiol proteases like papain or bromelain as catalysts,* the reaction mixture must contain a reductive agent such as β-mercaptoethanol, to prevent oxidation of the thiol function in the active site. Sometimes, it is more effective to carry out the reaction under argon or nitrogen atmosphere.

- *When the protease has affinity for both acyl-donor and nucleophile,* Cam esters can not be used as α-carboxyl protecting group of the nucleophile, because the risk of further reaction with the product formed. Then, Cam ester must be introduced once the product is synthesized. The preparation of Cam esters involves two steps: removal of terminal carboxyl ester, usually methyl or ethyl and incorporation of Cam ester (see Protocol 3, Z-Gly-Trp-OCam). The yield of this procedure decreased with the length of the peptide. Hence, other esters such as benzyl or allyl are also suitable to obtain a product than can be used directly in the next step (see Protocol 2, Z-Tyr-Gly-Gly-Phe-Leu-OEt, and Protocol 3, H-Gly-Trp-Met-Asp-Phe-NH$_2$).

5   Remarks and Conclusions

The reaction conditions presented in the protocol section are the result of optimization studies regarding high enzymatic activity and maximum yield. Valuable information is provided by systematic studies on dipeptide synthesis in organic media that served as starting reaction conditions. Thus, the following variables have to be modified for each particular enzymatic coupling: type of protease, support for enzyme deposition, enzyme loading onto the support, solvent, additives such as triethylamine, substrate concentration and structure, concentration of aqueous buffer or thermodynamic water activity. General comments on these variables and their effect on the reactions studied are given below.

5.1   Synthetic strategy and enzyme selection.

The synthesis of selected pentapeptides was carried out using a convergent strategy (Figures 4, 5 and 6): fragments were prepared by stepwise synthesis and then coupled together to yield the target peptide. The primary specificity of proteinases [27] provides the basis of the retrosynthetic analysis for the selection of fragments. However, the final success in devising a strategy will depend on the amino acid sequence of fragments. The conformation of the fragment or the amino acids nearby the bond that has to be formed may affect the enzyme-substrate interaction and, in turn, the reaction yield. For instance, the strategy for the preparation of Leu-enkephalin C-terminal ethyl ester derivative (Figure 5) was different from the one used for the C-terminal amide (Figure 4) due to the fact that the 4 + 1 fragment condensation failed [28].

The broad substrate specificity of some proteases like papain provides a useful tool for incorporation of non-coded synthetic amino acids. For instance, papain was highly efficient for the diastereoselective synthesis of dipeptides containing the non-coded amino acid γ-carboxyglutamic acid (Table 1) with enantiomeric excesses higher than 99.5% [29]. The optimization studies also include the search of new proteases to cover all the possibilities of peptide bond formation. In this sense, bromelain a thiolprotease with a substrate specificity similar to papain and not very much exploited for peptide synthesis [30], was used to catalyze the fragment condensation reaction (see Protocol 2, Z-Tyr-Gly-Gly-Phe-Leu-OEt). In this particular reaction bromelain was more efficient than papain (97% versus 52% yield) [28].

Peptide bond formation reactions catalyzed by serine and thiol proteases were performed under kinetically controlled conditions. The thermodynamically controlled approach was used in thermolysin-catalyzed reactions.

5.2   Selection of solvent

Solvent plays an important role in enzymatic peptide synthesis in organic media [6, 31]. Peptides are usually insoluble in apolar organic media such as hexane, diisopropyl ether, etc. Hence, solvents like dimethylsulfoxide, dimethylformamide, dioxane, acetonitrile and in some cases ethyl acetate and dichloromethane are needed to solubilize them efficiently. Previous fundamental studies on the influence of solvent in enzymatic dipeptide synthesis, showed that both acetonitrile and ethyl acetate were the best solvents providing high enzymatic activities and product yields [32, 35].

Ethyl acetate and, in particular, acetonitrile were both used to perform the synthesis of the examples presented. These solvents had good solubilizing properties for both oligopeptide fragments and starting amino acid derivatives. Furthermore, the activity and stability of the proteases employed were more than acceptable in these media.

5.3   Biocatalyst configuration

A biocatalyst configuration that has been shown to be quite useful in enzymatic peptide synthesis in organic media is the protease deposited onto a solid support [36]. Studies have been performed to identify the most suitable support material for proteases in peptide synthesis. Among them, the inorganic naturally occurring diatomaceous earth, Celite®, and synthetic porous polyamide gave the best results [37, 38]. The choice of one or another depended on the reaction considered. For instance, for the enzymatic synthesis of Leu-enkephalin derivatives, Celite was used as support material in all steps. The same for CCK-5 except for the final 3+2 enzymatic fragment condensation (see Protocol 3, H-Gly-Trp-Met-Asp-Phe-NH$_2$) where polyamide gave the best results. The papain-catalyzed synthesis of carboxyglutamic acid containing dipeptides was performed using papain adsorbed onto Celite, but similar results were obtained also with polyamide (unpublished results).

When enzymes are deposited onto solid supports the amount of enzyme load (enzyme loading) has to be optimized to avoid overloading and loss of enzyme and/or activity [39]. The optimum enzyme loading depends on the support, substrates and water concentration in the reaction medium. The accessible specific surface area of the support material controls the optimum values. In most cases, the best results were obtained for enzyme loadings ranging from 2 to 4 mg of enzyme/m^2. For higher enzyme loading, the activity reached a plateau or decreased, indicating mass transfer limitations [40, 41] and/or changes in the water distribution in the system. These two effects are obviously function of the substrate and water concentration.

The procedure used for drying the mixture enzyme-support also influences the activity of proteases deposited onto solid supports. In the reaction (Protocol 3, H-Gly-Trp-Met-Asp-Phe-NH$_2$), the catalytic activity of α-chymotrypsin deposited onto polyamide in acetonitrile exhibited hydration/dehydration hysteresis [42, 43]. Thus, depending on the way used to fix the hydration level of the enzyme-support preparation the final activity was different. The best results were obtained by equilibrating directly the α-chymotrypsin-polyamide mixture at constant water activity (a$_w$=0.1) without a previous removal of water under vacuum.

5.4   Water content/water activity

During protease-catalyzed peptide bond formation, water participates directly as a reaction product or indirectly generating hydrolysis products. Thermodynamically controlled peptide bond synthesis involves the formation of water. In kinetically controlled reactions the acyl-donor ester can be hydrolyzed and the peptide product may undergo secondary hydrolysis. Hence, the water concentration or the thermodynamic water activity is a key parameter in enzymatic peptide synthesis. Optimization of this variable is of vital importance: the water concentration must be kept as low as possible, but some water has to be present for enzyme activity [44]. The optimal water content (or water activity) has to be found for each particular coupling.

At low water concentrations the enzyme activity is low and has to be enhanced by modifications of substrate structure (see below) or by the use of additives such as triethylamine. Triethylamine was used in Protocol 1, Z-Tyr-Gly-OBzl and Z-Tyr-Gly-Gly-Phe-OMe, to neutralize *in situ* the hydrochloride or trifluoracetate salts of the amino group of the nucleophile. In α-chymotrypsin-catalyzed reactions (see Protocol 3, Z-Gly-Trp-Met-OAl, and H-Gly-Trp-Met-Asp-Phe-NH$_2$) addition of triethylamine was necessary even when the free base of the nucleophile was used. In presence of this organic base the enzymatic activity was around 2.5 times higher, reducing the reaction times by 3 to 6 times. Triethylamine could act as a water mimic solvent due to the existence of hydrogen bonding between TEA and the enzyme molecules [28, 41, 42]. This would improve the ionization state of the enzyme facilitating the enzyme substrate interaction [44]. Interestingly, this kind of activation was not observed for the rest of proteases used. For instance, in the reactions 1, 4, 6 and 7 of Table 1 catalyzed by papain onto Celite similar results were obtained with or without additions of triethylamine.

5.5   Substrate structure and concentration.

Substrate carboxyl component
The chemical nature of the ester moiety of the acyl-donor has a strong influence on enzyme activity and product yield in kinetically controlled synthesis [45]. When the reactivity of the enzyme towards the acyl-donor is high, the formation of by-products is greatly minimized during the reaction see Protocols 1, Z-Tyr-Gly-OBzl and 3, Z-Gly-Trp-Met-OAl [41]. Furthermore, the structure of the ester may influence the relative rates of hydrolysis and aminolysis of the acyl-enzyme complex.

In organic media, we observed that carboxamidomethyl esters (Cam) always gave the best results. For instance, in the last two reactions of Protocol 3, the use of Cam esters improved the product yield as well as the ratio synthesis/hydrolysis (S/H) compared with methyl or ethyl esters (see second last reaction of Protocol 3: 76% vs 50%, S/H 5.1 vs 3.1; last reaction of Protocol 3: 80% vs 70%, S/H 5.7 vs 3.4) [41, 42]. The reaction rates obtained with Cam esters were 4 to 10 times higher than with methyl or ethyl esters, reducing the reaction time by a factor of 6. Other esters that have been proved to be useful are allyl [42] (Protocol 3, H-Gly-Trp-Met-Asp-Phe-NH$_2$) and benzyl esters [41] (Protocol 1, Z-Tyr-Gly-Gly-Phe-OMe and 2, Z-Tyr-Gly-Gly-Phe-Leu-OEt) which gave results in between Cam and methyl or ethyl esters. Methyl and ethyl esters were in practice the first choice, in spite of their low reactivity, due to their simplicity and availability.

The N-α protecting group is also important in both thermodynamically and kinetically controlled approaches. It has been observed that the structure and physicochemical properties of the N-α group influence the enzymatic activity and reaction performance [46]. Working with α-chymotrypsin in organic media, hydrophilic α-amino protecting groups gave the highest activities. Moreover, amide type groups such as phenylacetamido (PhAc) have much more affinity for chymotrypsin than the urethane type like benzyloxycarbonyl. Phenylacetamido has also the advantage that it can be removed enzymatically by means of penicillin G acylase. This group and mandelylacetamido have been tested in our lab for the synthesis of oligopeptides using papain and chymotrypsin as catalyst providing both high enzymatic activities and yields [47]. Benzyloxycarbonyl (Z) and *tert*-butyloxycarbonyl (Boc) have the advantage that they are widely used in chemical peptide synthesis facilitating the combination of chemical and enzymatic steps. Furthermore, Z and Boc amino acid derivatives are readily commercially available.

Nucleophile structure

The C-α group of the nucleophile is also of paramount importance for enzymatic synthesis of oligopeptides. It is well-known that amino acid amides are good nucleophiles [48, 49]. However, when the product obtained is not the final target, it is of little interest from synthetic point of view. The terminal amide groups have to be removed using proteases with the risk of transpeptidation reactions [50] or with amidases not commercially available [51]. The use of amino acid esters (methyl, ethyl, benzyl) as nucleophiles has more synthetic applicability. Thus the product obtained can be used directly without any further chemical modification in the next enzymatic step or easily deprotected. It is important to select the C-α ester judiciously to avoid further reactions. When the protease has affinity for both the acyl donor and nucleophile, it is convenient to select a high reactive ester (i.e., Cam esters). Hence, side reactions with the nucleophile and product formed will be minimized.

Substrate concentration

Substrate concentration affects both enzymatic activity and yield in protease catalyzed peptide synthesis. Substrate or/and product inhibition must be identified in scaling up a particular enzymatic coupling. In test experiments a concentration of acyl-donor ranging from 20–30 mM was normally used with a 50% excess of nucleophile. For scaling up purposes, the effect of increasing the substrate concentration maintaining the molar ratio acyl-donor/nucleophile constant is usually investigated. In most of syntheses, the acyl donor is the limiting reactant. Moreover, an excess of the simplest substrate is used, independently of whether it was the acyl-donor or the nucleophile (Protocol 2, Z-Tyr-Gly-Gly-Phe-Leu-OEt). In enzymatic peptide synthesis it is possible to use the intermediates directly without purification. This is feasible because the main impurities, hydrolysis product of acyl-donor and excess of nucleophile, may not interfere in the next coupling. If one of these components interferes, the relative excesses of the acyl-donor or nucleophile can also be changed to consume completely one of the reactants.

Acknowledgments

This research was financially supported by the Comisión Interministerial para la Ciencia y la Tecnología C.I.C.Y.T. (SAF92–0261-C02 and BIO95–1083-CO2) and by the SCIENCE program of the European Union. The Department of Chemical Engineering UAB is a member of the Centre de Referència en Biotecnologia de la Generalitat de Catalunya.

Abbreviations

a_w :   Thermodynamic activity of water
ACN: acetonitrile
ATEE: N-acetyl-L-tyrosine ethyl ester
BAEE: N-α-benzoyl-L-arginine ethyl ester
Cam: carboxamidomethyl
EA: ethyl acetate
TEA: triethylamine
TEAP: Triethylamine-phosphoric acid buffer
TFA: trifluoroacetic acid
Z: Benzyloxycarbonyl
Boc: *tert*-butyloxycarbonyl
Gla: Carboxyglutamic acid
Bu^t: *tert*-butyl

References

1   Morihara K, Oka T, Tsuzuki H (1979) Semi-synthesis of human insulin by trypsin catalysed replacement of Ala-B30 by Thr in porcine insulin. *Nature* 280: 412–413

2   Isowa Y, Ohmori M, Ichikawa T et al. (1979) The thermolysin-catalyzed condensation reactions of N-substituted aspartic and glutamic acids with phenylalanine alkyl esters. *Tetrahedron Lett.* 28: 2611–2612

3   Andersen AJ, Fomsgaard J, Thorbek P, Aasmul-Olsen S (1991) Current possibilities in the technology of peptide synthesis. I Methods of synthesis and practical examples. *Chimica Oggi* 9(3): 17–24

4   Andersen AJ, Fomsgaard J, Thorbek P, Aasmul-Olsen S (1991) Current possibilities in the technology of peptide synth-

esis. II Process development in enzymatic peptide synthesis. *Chimica Oggi* 9 (4): 17–23

5   Jakubke H-D, Kuhl P, Könnecke A (1985) Basic principles of protease-catalyzed peptide bond formation. *Angew Chem Int Ed Engl* 24: 85–93

6   Dordick JS (1989) Enzymatic catalysis in monophasic organic solvents. *Enzyme Microb Technol* 11:194–211

7   Adlercreutz P, Mattiasson B (1987) Aspects of biocatalyst stability in organic solvents. *Biocatalysis* 1: 99–108.

8   Gupta MN (1992) Enzyme function in organic solvents. *Eur J Biochem* 203: 25–32

9   Feliu JA, de Mas C, López-Santín J (1995) Studies on papain action in the synthesis of Gly-Phe in two-liquid-phase media. *Enzyme Microb Technol* 17: 882–887

10 Mattiasson B, Adlercreutz P (1991) Tailoring the microenvironment of enzymes in water-poor systems. *TIBTECH* 9: 394–398

11 Klibanov AM (1986) Enzymes that work in organic solvents. *CHEMTECH* 16: 354–359

12 Wartchow CA, Callstrom MR, Bednarski, MD (1995) Stabilized proteases for peptide synthesis in acetonitrile solvent systems. In: CG Gebelein, E Carraher Jr (eds): *Industrial Biotechnology Polymers*. Technomic Lancaster, Pensylvania USA, 323–338

13 Bongers J, Heimer E (1994) Recent developments of enzymatic peptide synthesis. *Peptides* 15: 183–193.

14 Jakson DY, Burnier J, Quan C et al. (1994) A designed ligase for total synthesis of Ribonuclease A with unnatural catalytic residues. *Science* 266: 243–247

15 Olson GA, Olson RD, Kastin AJ, (1994) Endogenous Opiates: 1993. *Peptides* 15:1513–1556

16 Shively J, Reeve JR, Eysselein VE et al. (1987) Sequence analysis of a small cholecystokinin from canine brain and intestine. *Am J Physiol* 252:272–275

17 Hauschka PV, Lian JB, Cole DEC, Gundberg CM (1989) Osteocalcin and matrix gla protein: vitamin K-dependent proteins in bone. *Phys Rev* 69:99–1047

18 Esmon CT (1989) The roles of protein C and thrombomodulin in the regulation of blood coagulation. *J Biol Chem* 264: 4743–4746

19 Olivera BM, Rivier J, Clark C et al. (1990) diversity of conus neuropeptides. *Science* 249: 257–263

20 Halling P J (1994) Thermodynamic predictions for biocatalyst in nonconventional media: Theory, tests, and recomendations for experimental design and analysis. *Enzyme Microb Technol* 16: 178–206

21 Martinez J, Laur J, Castro B (1983) Carboxamidomethyl esters (CAM esters) as carboxyl protecting groups. *Tetrahedron Lett.* 24: 5219–5222

22 Bodanszky M (1984) Removal of urethane-type protecting groups. In K. Hafner, J M Lehn, C H Ress, P R Schleyer, B M Trost, R Zahradnik (eds): *Principles of peptide synthesis*. Springer-Verlag, Heidelberg, 98–102

23 Bodanszky M (1984) Removal of protecting groups. In K Hafner, JM Lehn, CH Ress, PR Schleyer, BM Trost, R Zahradnik (eds): *The practice of peptide synthesis*. Springer-Verlag, Heidelberg, 151–170.

24 Still WC, Kahn M, Mitra A (1978) Rapid Chromatography Technique for preparative separations with moderate resolution. *J Org Chem* 43: 2923–2925

25 Halling, PJ, Eichhorn, E, Kuhl, P, Jakubke, H-D (1995) Thermodynamics of solid-to-solid conversion and application to enzymic peptide synthesis. *Enzyme Microb Technol* 17: 601–606

26 Klein, JU, Cerovsky, V (1996) Protease-catalyzed synthesis of Leu-enkephalin in a solvent-free system. *Int J Peptide Protein Res* 47: 348–352

27 Morihara K (1987) Using proteases in peptide synthesis. *TIBTECH* 5:164–169.

28 Clapés P, Torres JL, Adlercreutz P (1995) Enzymatic peptide synthesis in low water content systems: preparative enzymatic synthesis of [Leu]- and [Met]-enkephalin derivatives. *Bioorgan Med Chem* 3: 245–255

29 Pera E, Torres JL, Clapés P (1995) Enzymatic synthesis of carboxyglutamic acid containing peptides in organic media. *Tetrahedron Lett* 37:3609–3612

30 Fruton JS (1982) Proteinase-catalyzed synthesis of peptide bonds. *Adv Enzymol Relat Areas Mol Biol* 53: 239–306

31 Cassells JM, Halling, PS (1989) Protease-catalyzed peptide synthesis in low water organic two phase systems and problems affecting it. *Biotechnol Bioeng* 33: 1489–1494

32 Kise H, Hayakawa A, Noritomi H (1990) Protease-catalyzed synthetic reactions and immobilization-activation of the enzymes in hydrophillic organic solvents. *J Biotechnol* 14: 239–254

33 Clapés P, Adlercreutz P, Mattiasson B (1990) Enzymatic peptide synthesis in organic media: nucleophile specificity and medium engineering in α-chymotrypsin catalyzed reactions. *Biotechnol Appl Biochem* 12: 376–386

34 Clapés P, Adlercreutz P, Mattiasson B (1990) Enzymatic peptide synthesis in organic media: a comparative study of water miscible and water inmiscible solvent systems. *Biotechnol Appl Biochem* 15: 323–338

35 Kise H, Fujimoto K, Noritomi H (1988) Enzymatic reactions in aqueous-organic media VI. Peptide synthesis by α-chymotrypsin in hydrophilic organic solvents. *J Biotechnol* 8: 279–290

36 Reslow M, Adlercreutz P, Mattiasson B (1988) On the importance of the support material for bioorganic synthesis. Influence of water partition between solvent,

enzyme and solid support in water-poor reaction media. *Eur J Biochem* 172: 572–578

37 Adlercreutz P (1991) On the importance of the support material for enzymatic synthesis in organic media. Support effects at controlled water activity. *Eur J Biochem* 119: 609–614

38 Capellas M, Benaiges MD, Caminal G et al. (1996) Influence of water activity and support material on the enzymatic synthesis of a CCK-8 tripeptide fragment. *Biocat Biotrans* 13: 165–178

39 Whetje E, Adlercreutz P, Mattiasson B (1993) Improved activity retention of enzymes deposited on solid supports. *Biotechnol Bioeng* 41: 171–178

40 Barros R, Whetje E, García FAP, Adlercreutz P (1998) Physical characterization of porous materials and correlation with the activity of immobilized enzyme in organic medium. *Biocat Biotrans* 16:67–85

41 Capellas M, Benaiges MD, Caminal G et al. (1996) Enzymatic synthesis of a CCK-8 tripeptide fragment in organic media. *Biotechnol Bioeng* 50: 700–708

42 Capellas M, Caminal G, Gonzalez G et al. (1997) Enzymatic condensation of cholecystokinin CCK-8 (4–6) and CCK-8 (7–8) peptide fragments in organic media. *Biotechnol Bioeng* 56: 456–463

43 Partridge J, Hutcheon GA, Moore BD, Halling PJ (1996) Exploiting hydration hysteresis for high activity of crosslinked subtilisin crystals in acetonitrile. *J Am. Chem Soc* 118: 1273–12877

44 Bell G, Halling P, Moore BD et al. (1995) Biocatalysis behaviour in low-water systems. *TIBTECH* 13: 468–473

45 Schellenberger, V, Görner A, Könnecke A, Jakubke H-D (1991) Protease-catalyzed peptide synthesis: prevention of side reactions in kinetically controlled reactions. *Peptide Res* 4: 265–269

46 Calvet S, Clapés P, Torres JL et al (1993) Enzymatic synthesis of X-Phe-Leu-NH$_2$ in low water content systems: influence of the N-αprotecting group and the reaction medium composition. *Biochim Biophys Acta* 1164: 189–196

47 Fité M, Alvaro G, Clapés P et al (1998) Reactivity and easily removable protecting groups for glycine in peptide synthesis using papain as catalayst. *Enzyme Microb Technol* 23: 199–203

48 Morihara K, Oka T (1977) α-chymotrypsin as the catalyst for peptide synthesis. *Biochem J* 163: 531–542

49 Nagashima T, Watanabe A, Kise H (1992) Peptide synthesis by proteases in organic solvents. Medium effect on substrate specificity. *Enzyme Microb Technol* 14: 842–847

50 Breddam K, Widmer F, Johansen JT (1980) Carboxypeptidase Y catalyzed transpeptidations and enzymatic peptide synthesis. *Carlsberg Res Commun* 45:237–247

51 Steinke D, Kula M-R (1990) Selective deamidation of peptide amides. *Angew Chem Int Ed Eng.* 29: 1139–1140

8 Enzyme Selectivity in Organic Media

Gianluca Ottolina and Sergio Riva

Contents

1 Introduction

For a long time biocatalytic transformations have been exclusively performed in homogeneous water solutions. More recently, to increase substrate solubility, biphasic systems (water plus water-immiscible organic solvents) have been used [1] and, even more recently, the so-called *nonaqueous enzymology* (enzymes suspended in homogeneous organic solvents) has been developed [2, 3]. The initial observation that enzymes are still active in nonaqueous media was made at the beginning of this century using tyrosinase [4], and these data were then revisited by Klibanov and co-workers at the beginning of the last decade [5].

The majority of the literature published on this subject in the last fifteen years describes the performances of hydrolases, but other classes of enzymes (lyases, dehydrogenases and oxidases) have been also studied. As has been pointed out in a previous chapter, performing enzymatic reactions in organic media can offer several advantages: better solubility of the substrates; easier

Methods and Tools in Biosciences and Medicine
Methods in non-aqueous enzymology, ed. by M. N. Gupta
© 2000 Birkhäuser Verlag Basel/Switzerland

reaction work-up; minimization of the thermoinactivation; suppression of possible side reaction; shift of the thermodynamic equilibrium (e.g. synthesis favoured over hydrolysis); possibility of tuning chemo-, regio- and enantioselectivity.

The previous chapters have also highlighted the parameters that influence enzymatic activity in organic solvents (enzyme form, water concentration, pH, nature of the organic solvent). The same parameters can also affect enzyme selectivity, but, in this case, the most remarkable effects occur by changing the organic solvent used. This aspect is very important, because tuning enzyme selectivity simply by changing the reaction medium gives new opportunity for biocatalysis. One of the goals of protein engineering is to modify enzyme selectivity at will by definite structural mutations of the proteins; the modification of enzyme selectivity by changing the nature of the organic solvent, the so-called *medium-engineering,* is even more attractive, because it might achieve the same results in a much simpler way.

2   Materials

Chemicals
The solvents used were at least 99% pure, and all chemicals were of analytical grade or purer.

Enzymes
- Polyphenol oxidase, horse liver alcohol dehydrogenase, mandelonitrile lyase and subtilisin were from Sigma.
- *Pseudomonas* sp. lipase was from Amano Pharmaceutical.

The assays that define the enzyme unit are described by the suppliers

3   Methods

3.1   Protocols

Protocol 1   General procedure for the pre-equilibration of the reagents

Reagents at defined water activity (a_w)

The reagents (enzymes, solvents, substrates) were separately placed in a beaker inside a sealed wide-mouth bottle containing a saturated aqueous solution of different salts (KCl, a_w=0.86; $Mg(NO_3)_2 \cdot 6H_2O$, a_w=0.53; $MgCl_2 \cdot 6H_2O$, a_w=0.38; LiCl, a_w=0.11) and incubated for at least two days [6,7].

Dry organic solvents

The organic solvents were distilled and then dehydrated by shaking with activated 4 Å molecular sieves, a_w<0.01 [6].

Protocol 2   General procedure for enzyme immobilization and lyophilization

Precipitation onto glass beads

Polyphenol oxidase: Non porous glass beads (~200 μm diameter, 1 g) were washed with 10% nitric acid for 30 min and then with distilled water until neutrality. Polyphenol oxidase (11.3 mg, 25,000 units expressed as tyrosinase activity) was dissolved in 50 mM phosphate buffer (pH 7) and the glass beads were added. The slurry was mixed and spread over a watch glass and left at room temperature for 5 hrs and then dried under vacuum.

Horse liver alcohol dehydrogenase (HLADH) : HLADH (20 mg, 30 units) was dissolved in 0.5 mL Tris-HCl buffer solution (20 mM, pH 7) containing 1 μg of NADH. Washed (as in the previous case) nonporous glass beads (75–150 μm, 1 g) were added, and the slurry was mixed and spread over a watch glass, left at room temperature for 5 hrs and then dried under vacuum.

Adsorption onto Celite

Polyphenol oxidase: Polyphenol oxidase (1.6 mg, 3,500 units expressed as tyrosinase activity) was dissolved in 50 mM phosphate buffer (pH 7) and deposited onto 1 g of Celite. The slurry was mixed and spread over a watch glass, then left at room temperature for 5 hrs and dried under vacuum.

Adsorption onto AVICEL

(R)-Mandelonitrile lyase: 2 g of microcristalline cellulose (AVICEL®, FMC Pharmaceutical Division) in 0.02 M acetate buffer (20 mL) were allowed to swell for 2 hrs. The solid was filtered and then 150 μL of a mandelonitrile lyase solution (0.02 M acetate buffer, pH 5.4, 700 units/mL) were added.

Lyophilization of subtilisin

Subtilisin Carlsberg (5 mg/mL) was dissolved in potassium phosphate buffer (20 mM, pH 7.8) and lyophilized.

Protocol 3   Oxidation of 4-substituted-phenols to 4-substituted-*ortho*-qui-
nones

Method 3.1

Immobilized polyphenol oxidase (10 mg) was added to 9 mL of organic solvent
at room temperature. A phosphate buffer solution (50 mM, pH 7, 25 µL) was
added, the slurry was stirred for 2 min, and then 1 mL of a solution of 4-substi-
tuted-phenol (25 mM) in the same organic solvent was added. Aliquots were
withdrawn from the reaction mixture at different times, and their absorbance
monitored at 395 nm.

Method 3.2

The immobilized polyphenol oxidase and the organic solvent were pre-equilibrated at a_w=0.86. The biocatalyst (10 mg) was added to 10 mL of a 25 mM solution of 4-substituted-phenol in the same organic solvent. Aliquots were withdrawn from the reaction mixture at different times, and their absorbance monitored at 395 nm.

Protocol 4   Formation of *(R)*-cyanohydrins from aldehydes

Organic solvent (20 mL saturated with 0.01 M acetate buffer, pH 5.4), 5 mmol
of aldehyde and 250 µL of HCN were added to 2 g of immobilized (*R*)-mandelo-
nitrile lyase. The suspension was shaken at 25 °C, and at different times aliquots were withdrawn from the reaction mixture and derivatized with (*R*)-(+)-
α-methoxy-α-trifluoromethyl phenylacetoyl chloride [8, 9] to be monitored by
GC.

Protocol 5   Reduction of 2-methylvaleraldehyde

2.5 mmol of 2-methyl-valeraldehyde, 10 mmol of ethanol, HLADH adsorbed on
glass beads (1 g), and 20 µL of aqueous buffer were added to 10 mL of ethyl
acetate presaturated with aqueous buffer (TrisHCl 20 mM, pH 7). The suspension was shaken at 25 °C and aliquots (1 µL) were withdrawn at different times
(over a period of 7 days) from the reaction mixture and monitored by GC (5 m
HP capillary column, 530 µm fused silica gel, N_2 carrier gas, 30 mL/min, detector and injector port temperature 250 °C).

Protocol 6   Transesterification of *sec*-phenethyl alcohol

Method 6.1

Dry organic solvent (1 mL), 0.01–0.07 mmol of enantiopure *sec*-phenethyl alcohol, 0.2 mmol of vinyl butyrate and 1 mg of subtilisin Carlsberg preequilibrated at a_w=0.11 were placed in a screw-capped vial. The vial was closed and shaken at 300 rpm, and at 45 °C. Periodically, aliquots (0.5 µL) were withdrawn (over a period of few hrs) and assayed by GC (10 m HP-5 capillary column coated with 5% phenyl/ 95% methyl silicone gum).

Method 6.2

Dry organic solvent (1 mL), 0.01 mmol of racemic *sec*-phenethyl alcohol, 0.2 mmol of vinyl butyrate, 50 mg of activated molecular sieves and 10 mg of subtilisin Carlsberg were placed in a screw-capped vial. The vial was closed and shaken at 180 rpm, and at 45 °C. Periodically, aliquots (1 µL) were withdrawn (over a period of few hrs) and the degree of conversion and the enantiomeric excess of the product determined by chiral GC (50 m CP-Cyclodextrin-β 2,3,6-M-19 from Chrompack, H_2 carrier gas, detector and injector port temperature 280 °C, oven temperature from 105 °C, initial time 30 min, to 140 °C with heating time of 0.5 °C/min).

Protocol 7   Transesterification of 2-(1-naphthoylamino)trimethylene dibutyrate

Hydrated organic solvent (1 mL, water content 0.01–1% v/v by mixing dry organic solvent and water), 2-(1-naphthoylamino) trimethylene dibutyrate (0.01 mmol), vinyl butyrate (0.2 mmol) and *Pseudomonas* sp. lipase powder (5–20 mg) were placed in a screw-capped vial. The vial was closed and shaken at 300 rpm and at 45 °C. Periodically, aliquots (20 µL) were withdrawn, centrifugate and the supernatant (5 µL) assayed by HPLC (Pirkle ionic D-phenylglicine chiral column from Regis Technology, eluent hexane/2-propanol 9/1, flow rate 1 mL/min, UV 280 nm).

4   Troubleshooting

There may not be a universally applicable theory [22] but the influence of the nature of the organic solvents on lipase and protease selectivity is a well accepted phenomenon, observed and demonstrated by several independent groups. As a matter of fact, up to now screening is still the best way to evaluate the influence of organic solvents on the selectivity of a specific enzymatic transformation, since the experiments are generally easy to handle and fast to be analyzed.

5   Remarks and Conclusions

The regioselective oxidation of *para*-substituted phenols catalyzed by polyphenol oxidase in chloroform was one of the first enzymatic reactions performed in organic solvents (Scheme 1) [5].

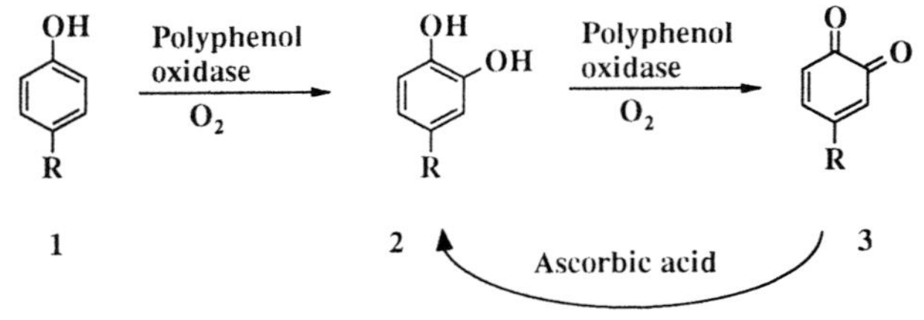

Scheme 1   Oxidation of para-substituted phenols catalyzed by polyphenol oxidase.

A large number of phenol derivatives (1) were oxidized to catechols (2) and then to *o*-quinones (3), which can be easily chemically reduced back to the more stable catechols (2) by an aqueous solution of ascorbic acid. The reaction was studied in several organic solvents (chloroform, toluene, *iso*-propyl ether, etc.) [10–12] showing a substrate specificity similar to that in water. This is a good example of a selective enzymatic transformation that can not be obtained in aqueous solution, because of the chemical instability of the products in water (Protocol 3).

Mandelonitrile lyase (oxynitrilase) from almond is an enzyme able to catalyze the carbon-carbon bond formation by addition of HCN to an aldehyde or to a ketone. According to Scheme 2, this reaction has been applied to the synthesis of optically active cyanohydrins starting from a great variety of aldehydes [13–15].

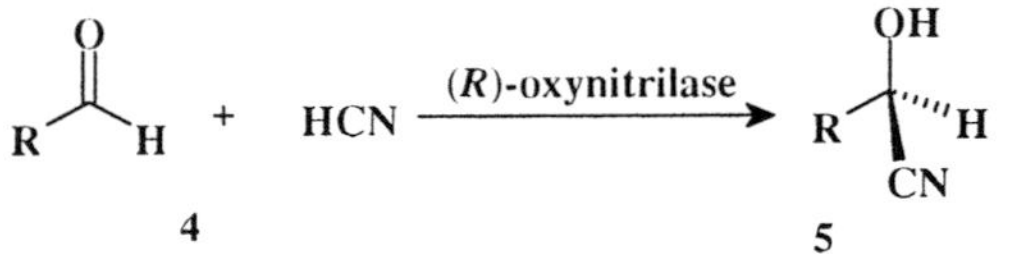

Scheme 2   Preparation of (*R*)-cyanohydrins by mandelonitrile lyase catalyzed addition of HCN to aldehydes.

Table 1   Solvent dependence of mandelonitrile lyase in the synthesis of *(R)*-5 cyanohydrins by addition of HCN to aldehydes 4 [11].

R	EtOAc			$i\mathrm{Pr_2O}$		
	t (hr)	Yield (%)	ee (%)	t (hr)	Yield (%)	ee (%)
C_6H_5	2.5	95	99	3	96	>99
3-Pyridyl	4.5	89	14	3	97	82
$H_3CS(CH_2)_2$	6.5	97	80	16	98	96
$(H_3C)_3C$	4.5	78	73	4.5	84	83

In water, the chemical addition competes with the enzymatic catalysis, giving products with low enantiomeric excesses. The chemical addition is partially suppressed in organic solvents, and the products obtained showed higher enantiomeric excesses. Table 1 summarizes the results obtained in the enzymatic synthesis of chiral cyanohydrines in two different organic solvents. The enantiomeric excesses of the products were adequately high in both solvents, but better results were obtained using *iso*-propyl ether (Protocol 4).

Horse liver alcohol dehydrogenase, a cofactor-dependent (NAD(H)) catalyst, is another enzyme whose performance in organic solvents has been studied [16, 17]. This approach was also investigated, among other things, to overcome the intrinsic instability of the cofactor in water. The highly polar nicotinamide cofactor is nearly insoluble in organic solvents, therefore to favor its fitting in the active site of the enzyme, it was co-precipitated with the catalyst from an aqueous buffer solution onto a solid support. This methodology can be generalized, in principle, for all the cofactor-dependent enzymes, its only drawback being the need of using the same enzyme to catalyze both the formation of the product and the regeneration of the cofactor, thus preventing the use of more efficient regenerating systems. Racemic aldehydes and ketones, such as 2-methyl-valeraldehyde, 2-phenyl-propionaldehyde and 2-chloro-cyclohexanone, have been reduced to the corresponding alcohols in organic solvents by means of HLADH (Protocol 5, Scheme 3, top) [16].

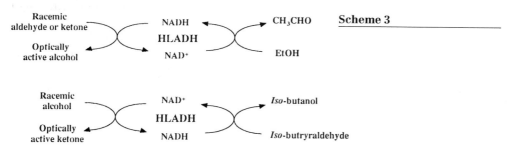

Racemic alcohols, such as *trans*-3-methyl-cyclohexanol and *cis*-2-methyl-cyclopentanol, have been oxidized to the corresponding ketones (Scheme 3, bottom). Different organic solvents were employed for these transformations (*iso*-propyl ether, butyl acetate and chloroform) but the ee of the products were always very high ((95%) [16].

The examples discussed so far have in common the large amount of water present in the reaction mixtures, an amount close to the saturation level of the organic solvents. Even though these systems can resemble biphasic systems, the enzymes were still active when the water content was accurately measured and kept below the saturation level. [12, 14, 17] In these experiments the nature of the solvents did not significantly influence enzyme selectivity. However, they were mainly focused on the demonstration that a stereoselective catalysis is possible even in organic solvents, by action of enzymes belonging to different classes.

On the other hand, the specific influence of the nature of the organic solvent on enzyme selectivity has been extensively investigated with lipases and proteases. A good example of inversion of chemoselectivity caused by the nature of the organic solvent is given by the transesterification of N-α-benzoyl-L-lysinol (6) (Scheme 4) with trifluoroethyl butyrate catalyzed by porcine pancreatic lipase [18].

Table 2   Solvent dependence of chemoselectivity of porcine pancreatic lipase in the acylation of 6.

Solvent	Initial rate, v (mM/h mg enz)		Chemoselectivity
	O-acylation, v_O	N-acylation, v_N	v_O/v_N
tert-butyl alcohol	0.71	0.30	2.4
tert-amyl alcohol	0.89	0.42	2.1
tert-butyl acetate	21	23	0.91
dichloromethane	1.4	1.6	0.88
acetonitrile	21	25	0.84
tetrahydrofuran	7.3	10	0.73
pyridine	3.5	4.8	0.73
1,2-dichloroethane	0.76	1.2	0.63
1,4-dioxane	5.9	10	0.59
tert-butyl methyl ether	58	110	0.53
nitrobenzene	1.6	3.5	0.46

Enzyme, 3.3–22 mg; solvent, 1 mL; trifluoroethyl butyrate, 30 mM; N-α-benzoyl-L-lysinol, 3 mM; 45 °C [16].

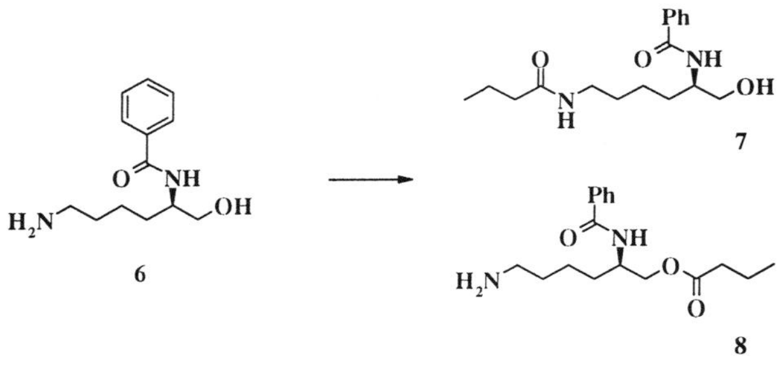

Scheme 4

The chemoselectivity (Table 2), expressed as the ratio of the initial rate for the O-acylation over the N-acylation, changed from 2.4 in *tert*-butyl alcohol to 0.46 in nitrobenzene, which means that crude porcine pancreatic lipase prefers to acylate the hydroxyl group of (6) in *tert*-butyl alcohol and the amino group of (6) in nitrobenzene. These results were correlated with different physicochemical properties of the solvents, and it was found that the best fitting was with the so-called *solvent acceptor number*, which reflects the ability of the solvent to serve as an acceptor of an electron pair [18].

Solvents can also affect lipases regioselectivity, as exemplified in the transesterification of 1,4-dibutyryloxy-2-octylbenzene (9). Butanolysis of (9) catalyzed by different lipases gave the two regioisomers 4-butyryloxy-2-octyl phenol (10) and 4-butyryloxy-3-octyl phenol (11) in the ratio reported in Table 3 (Scheme 5) [19].

Table 3   Solvent dependence of the regioselectivity of different lipases in the transesterification of 9 with butanol [17].

Lipase	Toluene $v_{2\text{-octyl}}/v_{3\text{-octyl}}$	Acetonitrile $v_{2\text{-octyl}}/v_{3\text{-octyl}}$
Cromobacterium viscosum	10	10
Candida cylindracea	10	10
Aspergillus niger	10	10
Pseudomonas cepacia	2.0	0.5

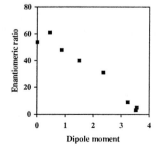

Scheme 5

As clearly indicated in Table 3, there was no effect of the organic solvent on the regioselectivity with the first three lipases. On the other hand, using *Pseudomonas cepacia* lipase the regioselectivity was inverted moving from toluene to acetonitrile. Other solvents were subsequently tested, and it was possible to correlate the observed regioselectivity to the logP of these solvents [19].

The effect of the solvents on the enantioselectivity (defined as the ratio of the specificity constants, (k_{cat}/K_M) of the two enantiomers, and expressed as enantiomeric ratio (E)) has been extensively studied [20]. The transesterification of *sec*-phenethyl alcohol with vinyl butyrate catalyzed by subtilisin showed a clear dependence on the physicochemical properties of the solvent employed (Protocol 6) [21, 22]. Figure 1 depicts a linear correlation of the enantiomeric ratio with the dipole moment or with the dielectric constant of the solvents used [21].

Solvents also affected significantly the prochiral selectivity of *Pseudomonas* sp. lipase [23], the prochiral selectivity being defined as the ratio of the rate of the accumulation of the major enantiomer over that of the minor enantiomer. The hydrolysis of the prochiral diester 2-(1 naphthoyl amino) trimethylene di-

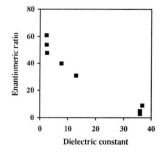

Figure 1   Enantioselectivity of subtilisin in the transesterification of sec-phenethyl alcohol with vinyl butyrate as a function of the dipole moment (left) or of the dielectric constant (right) of the solvent.

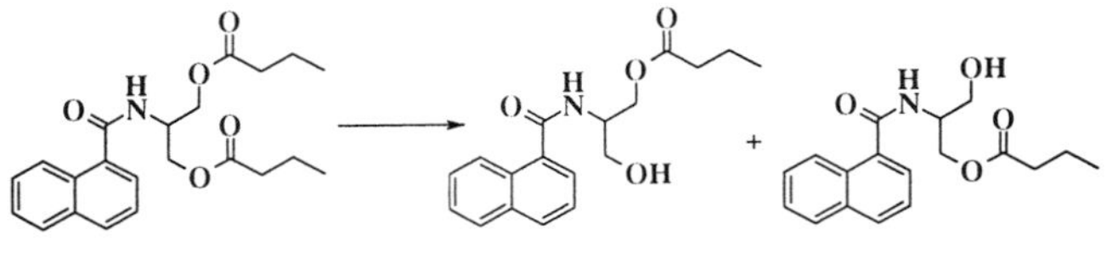

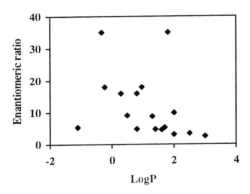

Scheme 6

12   13   14

butyrate (12) catalyzed by *Pseudomonas* sp. lipase in organic solvent is described in Scheme 6 (Protocol 7).

In this case, a correlation of the prochiral selectivity observed with the logP of the approximately 20 solvents used was not found, as shown in Figure 2.

Figure 2 Prochiral selectivity of *Pseudomonas* sp. lipase in the hydrolysis of 12 as a function of the hydrophobicity (logP) of the solvent.

In some cases, it has been possible even to invert enzyme enantiopreference by changing the solvent. For instance, the enantiomeric ratios for the transesterification of the chloroethyl ester of L- and D-N-acetyl-phenylalanine with propanol-catalyzed by *Aspergillus oryzae* protease are reported in Table 4 [24].

Table 4 Solvent inversion of regioselectivity of *Aspergillus oryzae* protease in the transesterification of N-acetyl-phenylalanine chloroethyl ester with propanol [22].

Solvent	Enantiomeric ratio (L/D)	Solvent	Enantiomeric ratio (L/D)
Acetonitrile	7.1	3-Octanone	0.73
Dimethylformamide	5.7	Nitrobenzene	0.60
Pyridine	4.3	*tert*-Butyl acetate	0.44
tert-Butyl alcohol	1.7	Triethylamine	0.34
Dioxane	1.3	Methyl-*tert*-butyl ether	0.34
Acetone	1.3	Cyclohexane	0.27
Tetrahydrofuran	1.3	Toluene	0.26
Cyclohexanone	1.1	Octane	0.24
Dichloromethane	0.88	Tetrachlorometane	0.19

The inversion of enantioselectivity was also noted with lipases. For instance, the hydrolysis of several prochiral 1,4 dihydropyridine derivatives (15) catalyzed by lipase AH in organic solvent gave the (S)-enantiomers (ee ~90%) in *iso*-propyl ether and the (R)-enantiomers (ee ~90%) in cyclohexane (Scheme 7) [25].

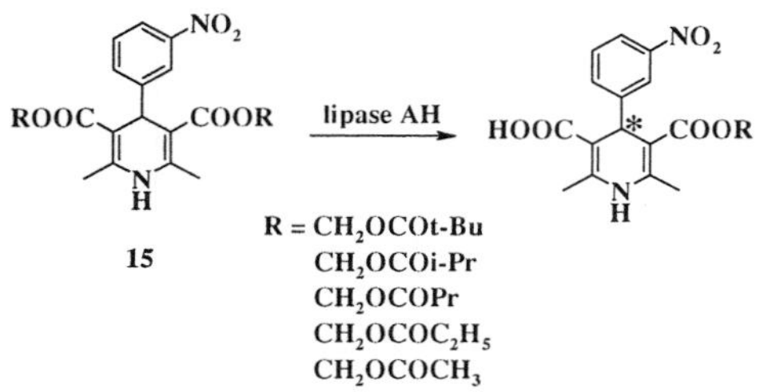

Scheme 7

$$\text{ROOC} \quad \text{COOR} \xrightarrow{\text{lipase AH}} \text{HOOC} \quad \text{COOR}$$

R = CH$_2$OCOt-Bu
CH$_2$OCOi-Pr
CH$_2$OCOPr
CH$_2$OCOC$_2$H$_5$
CH$_2$OCOCH$_3$

15

The inversion of enantioselectivity by changing the organic solvent was also reported in the esterification of 2-(4-chlorophenoxy) propionic acid with butanol catalyzed by *Candida cylindracea* lipase. Two groups found independently that this enzyme exhibits a preference for the R-enantiomer in hydrophobic solvents, a preference that can be reversed to the S-enantiomer in polar solvents [26, 27]. Surprisingly, the enantioselectivity in the buffer is not similar to that obtained in the polar solvents but to the one obtained in hydrophobic solvents.

These examples demonstrate that the influence of the solvent can be the determinant for the outcome of the enzymatic reaction. Many authors have tried to correlate the observed enzyme selectivity with macroscopic physicochemical properties of the solvents, like logP, dielectric constant, dipole moment, etc. It is not possible to give a general rule that correlates any of these parameters with the selectivity observed in each of the transformations reported in this chapter. This lack of correlation is exemplified in Figure 3. A systematic investigation on the esterification of unrelated compounds in different solvents catalyzed by lipase PS showed that the effects of the same solvent on the enantioselectivity is different with different substrates. It can be see that the best solvent for a certain substrate can be the worst for another one [28].

Besides the simple search for a correlation with physicochemical parameters of the solvents, three general theories have been proposed to explain the influence of organic media on enzyme selectivity. The first one assumes that the selectivity can be altered if the solvent molecules bind within the active site, therefore changing the interaction between enzyme and substrate [25, 29, 30]. Another theory suggests that the solvent could modify the enzyme conformation, altering in this way the molecular recognition process between substrate and enzyme [21, 26, 27]. The third theory suggests a dependence of selectivity on the energetics of substrate solvation [31, 32].

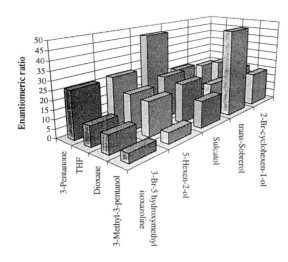

Figure 3 Enantioselectivity of lipase PS for various substrates in various solvents. In the case of *trans*-sobrerol the plotted values are $1/10^{th}$ of the real ones.

Acknowledgments

We thank the Biotechnology Programme of the European Commission (BI04-CT95–0231) and the CNR Target Project on Biotechnology for financial support.

References

1   Carrea G (1984) Biocatalysis in water-organic solvent two-phase systems. *Trends Biotechnol* 2: 102–106

2   J Tramper, MH Vermüe, HH Beeftink, U von Stockar (eds) (1992) *Biocatalysis in non conventional media. Progress in biotechnology*. Volume 8. Elsevier, Amsterdam, The Netherlands.

3   AMP Koskinen, AM Klibanov (eds) (1996) *Enzymatic reactions in organic media*. Blackie Academic & Professional, London

4   Bourquelot E, Bertrand G (1895) Le bleuissement et le noircissiment des champignons *Compt Rend Soc Biol* 47: 582–584

5   Kazandjian RZ, Klibanov AM (1985) Regioselective oxidation of phenols catalyzed by polyphenol oxidase in chloroform. *J Am Chem Soc* 107: 5448–5459

6   Valivety RH, Halling PJ, Macrea AR (1992) Reaction rate with suspended lipase catalyst shows similar dependence on water activity in different organic solvents. *Biochim Biophys Acta* 1118: 218–222

7   Greenspan L (1977) Humidity fixed points of binary saturated aqueous solutions. *J Res Natl Bur Stand A Phys Chem* 81A: 89–96

8   Dale JA, Dull DL, Mosher HS (1969) α-Methoxy-α-trifluoromethylphenylacetic acid, a versatile reagent for the determination of enantiomeric composition of alcohols and amines. *J Org Chem* 34: 2543–2549

9   Eliott JD, Choi VMF, Johnson WS (1983) Asymmetric synthesis via acetal templates. 5. Reactions with cyanotrimethylsilane. Enantioselective preparation of cyanohydrins and derivatives. *J Org Chem* 48: 2294–2295

10 Estrada P, Sanchez-Muñiz R, Acebal C et al. (1991) Characterization and optimization of immobilized polyphenol oxidase in low-water organic solvents. *Biotechnol Appl Biochem* 14: 12–20

11 Burton SG, Ducan JR, Kaye PT, Rose PD (1993) Activity of mushroom polyphenol oxidase in organic medium. *Biotechnol Bioeng* 42: 938–944

12 Yang Z, Robb DA (1994) Partition coefficients of substrates and products and solvent selection for biocatalysis under nearly anhydrous conditions. *Biotechnol Bioeng* 43: 365–370

23 Effenberger F, Ziegler T, Förster S (1987) Enzyme-catalyzed cyanohydrin synthesis in organic solvents. *Angew Chem Int Ed Engl* 26: 458–460

14 Wehtje E, Adlercreutz P, Mattiasson B (1990) Formation of C–C bonds by mandelonitrile lyase in organic solvents. *Biotechnol Bioeng* 36: 39–46

15 Effenberger F (1994) Synthesis and reactions of optically active cyanohydrins. *Angew Chem Int Ed Engl* 33: 1555–1564

16 Grunwald J, Wirz B, Scollar MP, Klibanov AM (1986) Asymmetric oxidoreductions catalyzed by alcohol dehydrogenase in organic solvents. *J.Am.Chem.Soc.* 108: 6732–6734

17 Virto C, Svensson I, Adlercreutz P, Mattiasson B (1995) Catalytic activity of noncovalent complexes of horse liver alcohol dehydrogenase, NAD+ and polymers, dissolved or suspended in organic solvents. *Biotechnology Lett* 17: 877–882

18 Tawaki S, Klibanov AM (1993) Chemoselectivity of enzyme in anhydrous media is strongly solvent dependent. *Biocatalysis* 8: 3–19

19 Rubio E, Klibanov AM (1991) Effect of the solvent on enzyme regioselectivity. *J Am Chem Soc* 113: 695–696

20 Carrea G, Ottolina G, Riva S (1995) Role of the solvents in the control of enzyme selectivity in organic media. *Trends Biotechnol* 13: 63–70

21 Fitzpatrick PA, Klibanov AM (1991) How can the solvent affect enzyme enantioselctivity? *J Am Chem Soc* 113: 3166–3171

22 Colombo G, Ottolina G, Carrea G et al. (1998) Application of structure-based thermodynamic calculations to the rationalization of the enantioselectivity of subtilisin in organic solvents. *Tetrahedron Asymmetry* 9: 1205–1214

23 Terradas F, Teston-Henry M, Fitzpatrick PA, Klibanov AM (1993) Marked dependence of enzyme prochiral selectivity on the solvent. *J Am Chem Soc* 115: 390–396

24 Tawaki S, Klibanov AM (1992) Inversion of enzyme enantioselectivity mediated by solvent. *J Am Chem Soc* 114: 1882–1884

25 Hirose Y, Kaiya K, Sasaki I et al. (1992) Drastic solvent effect on lipase-catalyzed enantioselective hydrolysis of prochiral 1,4-dihydropyridines. *Tetrahedron Lett* 33: 7157–7160

26 Wu HS, Chu FY, Wang K (1991) Reversible enantioselcectivity of enzymatic reactions by media. *Bioorg Med Chem Lett* 1: 339–342

27 Ueji S, Fujio R, Okubo N et al. (1992) Solvent-induced inversion of enantioselectivity in lipase-catalyzed esterification of 2-phenoxypropionic acids. *Biotechnol Lett* 14: 163–168

28 Carrea G, Ottolina G, Riva S, Secundo F (1992) Effect of reaction conditions on the activity and enantioselectivity of lipases in organic solvents. In: J Tramper, MH Vermüe, HH Beeftink, U von Stockar (eds). *Biocatalysis in non conventional media. Progress in biotechnology.* Volume 8. P. 111–119 Elsevier, Amsterdam, The Netherlands

29 Nakamura K, Takebe J, Kitayama T, Ohno A (1991) Effect of solvent structure on enantioselectivity of lipase-catalyzed transesterification. *Tetrahedron Lett* 32: 4941–4944

30 Secundo F, Riva S, Carrea G (1992) Effects of medium and of reaction conditions on the enantioseleivity of lipases in organic solvents and possible rationales. *Tetrahedron Asymmetry* 3: 267–280

31 Ke T, Wescott CR, Klibanov AM (1996) Prediction of the solvent dependence of enzymatic prochiral selectivity by means of structure-based thermodynamic calculations. *J Am Chem Soc* 118: 3366–3374

32 Wescott CR, Noritomi H, Klibanov AM (1996) Rational control of enzymatic enantioselectivity trough solvation thermodynamics. *J Am Chem Soc* 118: 10365–10370

9 Sugar Transformations Using Enzymes in Non-Aqueous Media

Sergio Riva and Gabriella Roda

Contents

1 Introduction

The peculiar properties of enzymes, especially high selectivity and mild reaction conditions, have boosted their popularity among organic chemists. These biocatalysts appear to be particularly suitable to support natural products synthesis and, more specifically, carbohydrate modification [1]. In fact, the presence of several stereogenic carbons as well as the similar chemical reactivity of the numerous hydroxy groups of the sugar molecules still make the transformation of these compounds a challenging target.

The application of enzymatic catalysis in this area has followed three main pathways:

Methods and Tools in Biosciences and Medicine
Methods in non-aqueous enzymology, ed. by M. N. Gupta
© 2000 Birkhäuser Verlag Basel/Switzerland

- Specific enzymes (aldolases, kinases, synthetases) have been employed for the synthesis of sugars and sugar derivatives from simpler precursors.
- Specific sugar transferases and sugar hydrolases (glycosidases) have been used to transform suitable glycosides into valuable di- and oligosaccharides as well as into their corresponding conjugates with other hydroxylated compounds.
- Hydrolytic enzymes such as lipases and proteases have found interesting applications in the regioselective acylation and deacylation of carbohydrates. The unusual exploitation of these enzymes (lipids and proteins are the natural substrates of lipases and proteases, respectively) in organic solvents constitutes an interesting approach to the selective protection of the different sugar hydroxyl groups. Besides that, carbohydrate esters are interesting compounds with potential applications as biodegradable surfactants, antitumoral agents and plant growth inhibitors [2].

Twelve years ago, Klibanov and co-workers reported the use of lipases for the regioselective esterification of primary [3] or secondary [4] hydroxyls of monosaccharides, and, later on, the use of the protease subtilisin for the modification of primary hydroxyl groups of di- and oligosaccharides as well as of natural glycosides [5]. The molecules acylated in this work (depicted in Figure 1) illustrate the versatility of this approach and, in a way, these compounds can be considered the *natural ancestors* of the numerous applications

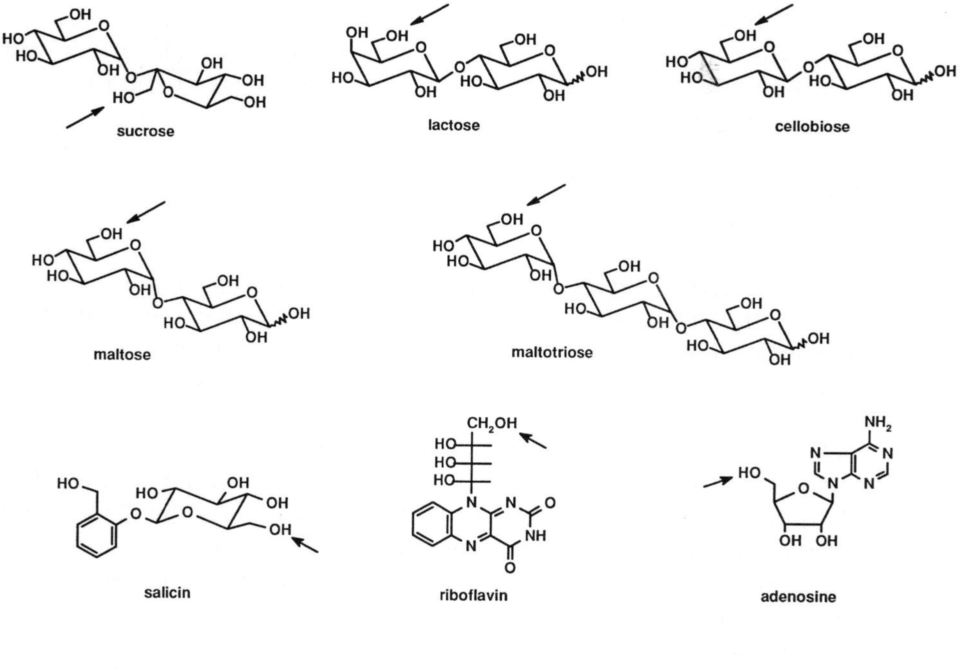

Figure 1 Acylation of di- and trisaccharides and of natural glycosides catalyzed by subtilisin in DMF (from [5]).

that have been reported in the literature since then. As this subject has already been covered in a recent review [6], only few representative examples will be described and commented upon here. The protocols reported in the Methods section will present additional cases taken from the research performed at the Istituto di Biocatalisi and Riconoscimento Molecolare (IBRM).

Specifically, lipases and proteases have been exploited

- to increase the basic knowledge on selectivity of hydrolases. For instance, the rational control of enzyme-catalyzed regioselectivity has been studied using sucrose acylation by vinyl esters in organic media as a model [7].
- to catalyze a protective step in chemo-enzymatic syntheses of sugar derivatives. As an example, subtilisin-catalyzed esterification of lactosides has been used for a chemo-enzymatic approach to synthesis of 6′-deoxy-6′-fluoro- and 6-deoxy-6-fluoro-lactosides [8].
- to synthesize specific esters of natural glycosides. For instance, a chemo-enzymatic approach to the preparation of some phenylpropenoate esters (cinnamate, coumarate, feruloate) of the flavonoid isoquercitrin has been used to overcome the inability to introduce these acyl moieties directly, by transesterification with the corresponding activated esters [9].
- for the large-scale production of new biosurfactants. To give a specific approach, an enzyme-based solvent-free process has been developed for the acylation of sugar glycosides [10]. The reactions were carried out by simply mixing the starting sugar with a fatty acid in the presence of an immobilized lipase, the water generated in the acylation reaction being removed in vacuo at 70 °C. In a pilot reactor, the reaction was performed on a 20 kg scale using a mixture of α- and β-ethyl-D-glucopyranosides, which in turn were prepared by glycosylation in ethanol by ion-exchange resin catalysis, and different fatty acids. The reaction yields of 6-O-monoester were 85–90% within 24 hrs, and the enzyme was recycled several times with no noticeable loss of activity.
- for the large-scale production of sugar acrylates, which are suitable monomers for the preparation of new polymeric hydrogels. A two steps chemo-enzymatic synthetic strategy has been developed to prepare poly(sugar acrylate)s [11]. In the first step an enzyme has been used to catalyze the regioselective acryloylation of a sugar, then the derivatized sugar has been chemically polymerized (typically by a free radical mechanism) to give a long-chain, linear poly (sugar acrylate).

Glycosidases are another class of hydrolytic enzymes that can reverse their activity under suitable reaction conditions [12]. Hundreds of papers have been published on the attempted exploitation of these enzymes for the direct synthesis of di- and oligosaccharides. The biocatalysts have been used for the preparation of glycosides of simple alcohols in water-organic two-phase systems, in which the organic phase is the reacting alcohol. Yields were variable, depending on the ratio of water to alcohol, and on sugar concentration. This system has the advantage of providing a simple and inexpensive procedure to prepare synthetically useful glycosides avoiding protective and activating reac-

a) REVERSE HYDROLYSIS   <u>Scheme 1</u>

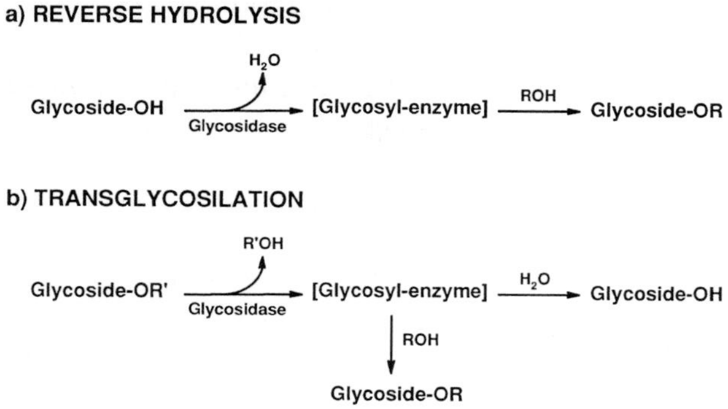

H_2O

Glycoside-OH ——————→ [Glycosyl-enzyme] ——ROH——→ Glycoside-OR
      Glycosidase

b) TRANSGLYCOSILATION

$R'OH$

Glycoside-OR' ——————→ [Glycosyl-enzyme] ——H_2O——→ Glycoside-OH
      Glycosidase

                      ↓ ROH

               Glycoside-OR

tions. For example, benzyl and allyl glycosides can be prepared on a multigram scale in a single step. Scheme 1 shows the two experimental approaches that have been used, the so-called "reverse hydrolysis" [13] and the transglycosylation procedure [14].

Glycosides bearing a spacer arm, which can be used to make glycoconjugates or to be coupled to a solid support for affinity chromatography can also be synthetized, as will be exemplified in one of the following sections.

2   Materials

Chemicals
- Vinyl acetate, trifluoroethyl butanoate, dibenzyl malonate, allyl alcohol, β-D-glucose, naringin, rutin and the solvents used were from Aldrich.
- Methyl 4,6-O-benzylidene-α- or β-D-glucopyranosides [15], benzyl β-lactose and 4-pentenyl trichloroethyl carbonate and trifluoroethyl levulinate [16], N-acryloyl-aminoethoxyethanol [17] and the benzyl amide of lactobionic acid were prepared according to standard procedures.

Enzymes
- Lipase PS (from Pseudomonas cepacia) was purchased from Amano Pharmaceutical Co.
- Novozym 435 (lipase B from Candida antarctica immobilzed on macroporous acrylic resin) was obtained from Novo-Nordisk.
- Subtilisin and β-glucosidase (from almonds) were from Sigma.

Equipment
- TLC were performed on Merck Silica Gel 60 F_{254} plates.
- Flash chromatographies were performed using Merck Silica Gel 60 (230–400 Mesh).

- NMR spectra were obtained using either a Bruker AC-300 (300 MHz) or a Bruker AC-200 (200 MHz).
- Reaction systems were shaken on a G24 environmental incubation shaker, New Brunswick Scientific Co.

3   Methods

3.1   Adsorption of lipase PS on celite

Lipase PS (3 g) was carefully mixed with Hyflo Super Cell celite (10 g, Fluka). Phosphate buffer 0.1 M pH 7 (10 ml) was added and, after shaking, the liquid in the slurry was allowed to evaporate at room temperature for 3 days, mixing the contents every 12 hrs.

3.2   Lyophilization and pH-adjustment of subtilisin

Subtilisin (Protease-type VIII, Sigma, 1 g) was dissolved in water (70 ml) containing 0.5 g of K_2HPO_4. The pH was adjusted to 7.4, the solution was freeze-dried and lyophilized.

3.3   Protocols

Protocol 1   Selective acylation of methyl 4,6-*O*-benzylidene-α- or β-D-gluco-pyranosides catalyzed by lipase PS [15]

Methyl 4,6-*O*-benzylidene-α-D-glucopyranoside (**1**, Figure 2, 200 mg) was dissolved in vinyl acetate (10 mL). Lipase PS adsorbed on celite (1 g) was added and the suspension was shaken at 45 °C and 250 rpm for 7 hrs. TLC analysis (hexane: EtOAc, 1:1) showed a 98% conversion to a single new product, which was isolated by filtration of the solid enzyme and evaporation of the solvent to give 220 mg of methyl 2-*O*-acetyl-4,6-*O*-benzylidene-α-D-glucopyranoside (**1a**).

Under similar conditions, methyl 4,6-*O*-benzylidene-β-D-glucopyranoside (**2**, Figure 2) showed a 98% conversion to a mixture of two monoesters in a 94:6 ratio after 1 hr. Usual work-up and purification by flash chromatography (hexane: EtOAc, 1:1) yielded 192 mg (86%) of methyl 3-*O*-acetyl-4,6-*O*-benzylidene-α-D-glucopyranoside (**2a**).

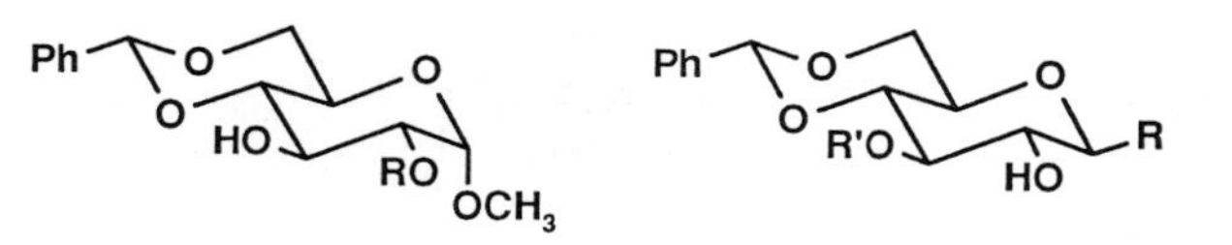

Figure 2   Formulae of compounds 1–4

1   R = H
1a R = Ac

2   R = OMe ; R' = H
2a R = OMe ; R' = Ac

3   R = SEt ; R' = H
3a R = SEt ; R' = Ac

4   R = SPh ; R' = H
4a R = SPh ; R' = Ac

Protocol 2   Enzymatic synthesis of benzyl 6′-*O*-(4-pentenyloxy)carbonyl-2′-*O*-levulinyl-β-lactose [16]

Benzyl β-lactose (**15**, Figure 4, 100 mg) was dissolved in tert-amyl alcohol (10 ml) containing 2 ml of 4-pentenyl trichloroethyl carbonate (**14**, Figure 3) and the solution was shaken at 45 °C for 7 days in the presence of 300 mg of lipase PS adsorbed on celite. The enzyme was filtered, the solvent evaporated and the crude residue purified by flash chromatography (eluent EtOAc: MeOH 10:0.8) to give 80 mg (66% yield) of benzyl 6′-*O*-(4-pentenyloxy)carbonyl-β-lactose (**15a**). This compound was then dissolved in tert-amyl alcohol (10 ml) containing 2 ml of trifluoroethyl levulinate (**12**, Figure 3), 300 mg of Novozym 435 were added and the suspension was shaken at 45 °C for 3 days. The enzyme was filtered, the solvent evaporated and the crude residue purified by flash chromatography (eluent EtOAc:hexane, 10:0.5) to give 75 mg (79% yield) of benzyl 6′-*O*-(4-pentenyloxy)carbonyl-2′-*O*-levuninyl-β-lactose (**15b**).

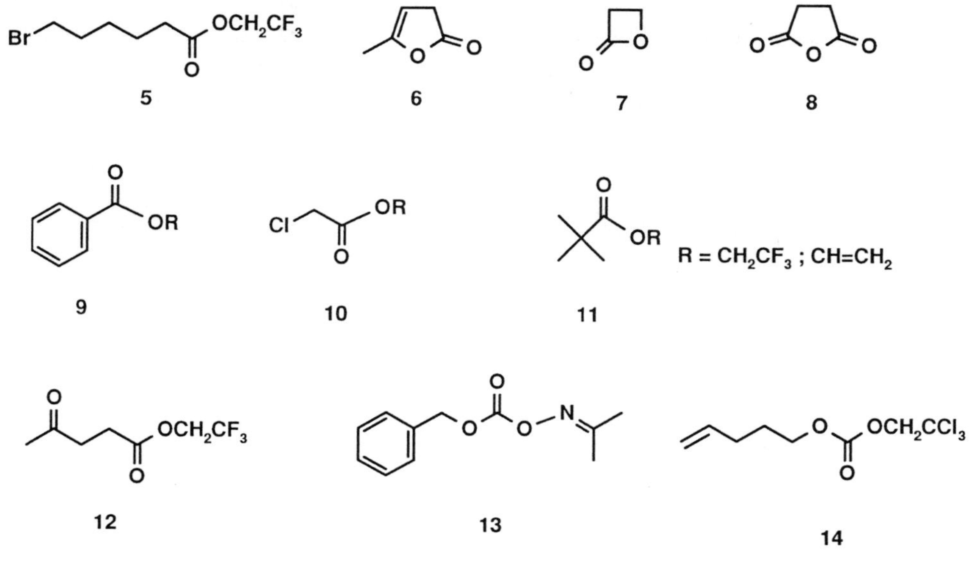

5

6

7

8

9

10

11

R = CH₂CF₃ ; CH=CH₂

12

13

14

Figure 3   Different activated esters that have been used as acyl donors (from [16, 23–25]).

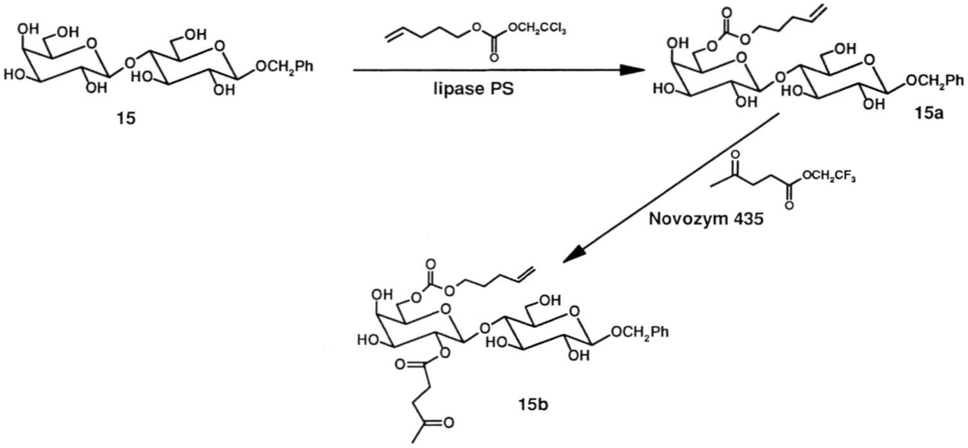
15
lipase PS
15a
Novozym 435
15b

Figure 4 Enzymatic synthesis of benzyl 6′-*O*-(4-pentenyloxy)carbonyl-2′-*O*-levulinyl-β-lactose (from [16]).

Protocol 3 Regioselective enzymatic acylation of the benzylamide of lactobionic acid

The benzyl amide **16** (Figure 5, 220 mg, 0.5 mmol) was dissolved in 9 ml a mixture of tert-amyl alcohol pyridine (2:1 v/v) and 2 ml of trifluoroethyl butanoate was added. Novozym 435 (200 mg) or lipase PS adsorbed on celite (500 mg) or subtilisin (50 mg) were added and the suspensions shaken at 45 °C for 48 hrs. The enzyme was filtered, the solvent evaporated and the residue purified by flash chromatography (eluent EtOAc:MeOH:H_2O, 10:2:0.5) to give, respectively, 110 mg (43% yield) of the 6-*O*-butanoate **16a** or 71 mg (28% yield) of the 6'-*O*-butanoate **16b** or 50 mg (17% yield) of the 6,6′-*O*-dibutanoate **16c** of the benzylamide of lactobionic acid.

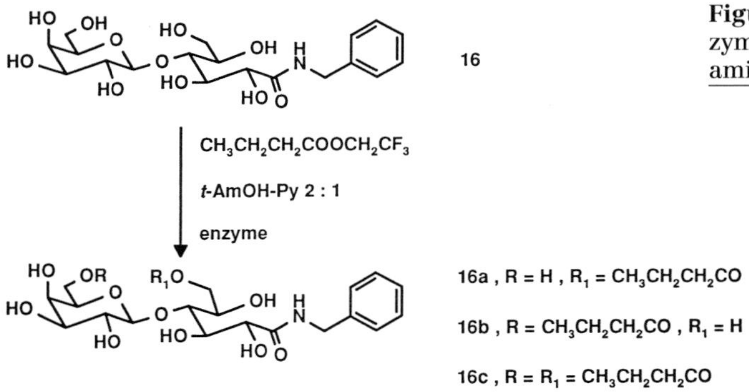
16
$CH_3CH_2CH_2COOCH_2CF_3$
t-AmOH-Py 2 : 1
enzyme
16a , R = H , R_1 = $CH_3CH_2CH_2CO$
16b , R = $CH_3CH_2CH_2CO$, R_1 = H
16c , R = R_1 = $CH_3CH_2CH_2CO$

Figure 5 Regioselective enzymatic acylation of the benzyl amide of lactobionic acid.

Protocol 4   Regioselective acetylation of rutin [18]

Rutin (**17**, Figure 6, 120 mg, 0.2 mmol) was dissolved in 0.5 ml of pyridine. Acetone (4.5 ml), vinyl acetate (0.2 ml) and Novozym 435 (150 mg) were added and the suspension shaken at 45 °C for 48 hrs, following the conversion by TLC (eluent $CHCl_3$:MeOH:H_2O, 8:2:0.3). The crude residue obtained after filtration of the enzyme and evaporation of the solvent was purified by flash chromatography to give 126 mg of 3",4'''-*O*-di-acetyl-rutin (**17a**, 91% yield).

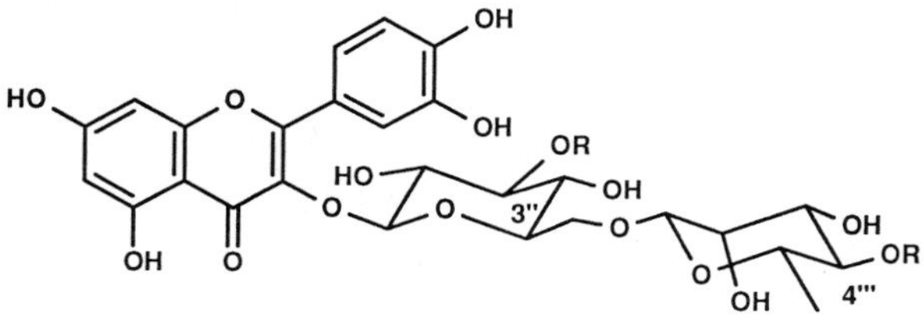

Figure 6   Formulae of compounds **17** and **17a**

Rutin (**17**) : R = H

3", 4'''-*O*-Acetyl-Rutin (**17a**) : R = Ac

Protocol 5   Chemoenzymatic synthesis of naringin 6"-*O*-malonate [19]

To a solution of the flavonol glycoside (**18**, Figure 7, 58 mg, 0.1 mmol) in anhydrous acetone containing 10% pyridine (3 ml), dibenzylmalonate (15 equivalents) and Novozym 435 (200 mg) were added and the suspension was shaken at 45 °C and 250 rpm for 10 days. After filtration of the enzyme and evaporation of the solvent, the residue was washed repeatedly with hexane and then purified by flash chromatography (eluent EtOAc-MeOH-H_2O, 90:10:3) to give

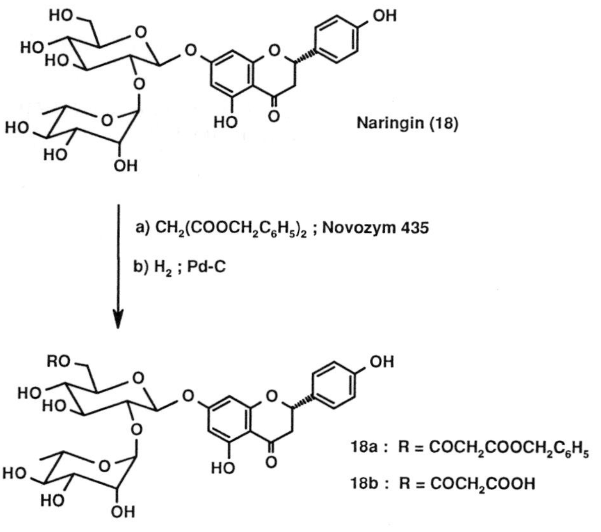

Figure 7   Chemoenzymatic synthesis of naringin 6"-*O*-malonate (from [18]).

Naringin (**18**)

a) $CH_2(COOCH_2C_6H_5)_2$; Novozym 435

b) H_2 ; Pd-C

18a : R = $COCH_2COOCH_2C_6H_5$

18b : R = $COCH_2COOH$

naringin 6″-*O*-benzylmalonate (**18a**) in 69% isolated yield (59 mg). A solution of this flavonol glycoside benzylmalonate (34 mg, 0.04 mmol) in 2 mL anhydrous THF with a catalytic amount of Pd/C (5%) was stirred for three days under H_2. The catalyst was filtered and solvent removed under vacuum at room temperature to give naringin 6″-*O*-malonate **18b** in quantitative yield.

Protocol 6   Enzymatic synthesis of allyl β-D-glucopyranoside [13].

Almond β-D-glucosidase (160 mg, 1100 U) was added to a solution of β-D-glucose (1.8 g, 10 mmol) in 3.5 mL of water and 30 mL of allyl alcohol. The mixture was stirred for 48 hrs at 50 °C, then was filtered to remove the precipitated enzyme. Following the evaporation of the solvent under reduced pressure, the crude residue was purified by flash chromatography (eluent CH_2Cl_2:MeOH:H_2O, 40:10:1) to give 1.18 g (60% yield) of allyl β-D-glucopyranoside (**19**, Figure 8).

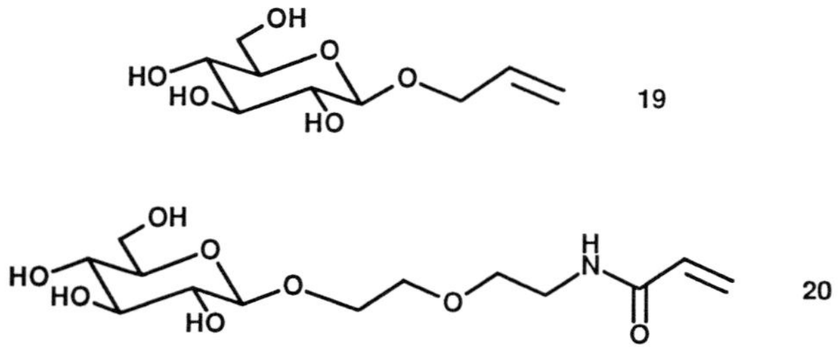

Figure 8   Formulae of compounds **19** and **20**

Protocol 7   Enzymatic synthesis of *N*-(acryloylaminoethoxy)ethyl-β-D-glucopyranoside [17]

A solution of β-D-glucose (1.8 g, 10 mmol) in 3.5 mL of water was added to a solution of *N*-acryloylaminoethoxyethanol (13.6 g, 8.5 mmol) and tert-butylhydroquinone (0.2 g) in 5 ml tert-butanol. β-D-Glucosidase from almond (360 mg, 2500 U) was added and the suspension was stirred for 48 hrs at 50 °C. When the reaction had reached its equilibrium, the mixture was filtered to remove the precipitated enzyme, the solvent was evaporated under reduced pressure, and the crude residue purified by flash chromatography (eluent CH_2Cl_2:MeOH, 8:2) to give 0.61 g (18% yield) of *N*-(acryloylaminoethoxy)ethyl-β-D-glucopyranoside (**20**, Figure 8).

4   Remarks and Conclusions

4.1   Hydrolases-catalyzed acylation and deacylation of sugars

The lipases and the proteases that have been used for the selective modification of sugars are listed in Table 1. However, most of the literature reports describe the performances of just few of them, and specifically the lipases from Pseudomonas or Candida species and the proteases from Bacillus species.

Table 1   Hydrolases used for the regioselective acylation of sugars in organic solvents.

Enzyme	Source	Trade name	Supplier[a]
Lipase	Aspergillus niger	Lipase AP6	Amano
	Candida antarctica	Novozym 435	Novo Nordisk
	Candida rugosa		Sigma
	Chromobacterium viscosum	Lipase CV	Finnsugar
	Humicola lanuginosa	Lipase CE5	Amano
	Mucor miehei	Lypozime	Novo Nordisk
	Pseudomonas cepacia	Lipase PS	Amano
	Porcine pancreas		Sigma
Protease	Bacillus amyloliquefasciens	Subtilisin BPN′ (Nagarase)	Sigma
	Bacillus licheniformis	Alkaline protease	Promega
	Bacillus licheniformis	Subtilisin Carlsberg	Sigma
	Bacillus licheniformis	Alcalase	Novo Nordisk
	Bacillus licheniformis	Optimase M-440	Solvay Enzyme
	Bacillus subtilis	Protease N	Amano
	Bacillus sp.	Proleather	Amano
	Streptomyces griseus	Protease Type XXI	Sigma
	Streptomyces sp.	Alkalophilic proteinase	Toyobo

a) The same enzyme might be commercialized with different trade names by other suppliers.

Protocols 1 and 2 exemplify the potential of this methodology as a protective step in a synthetic sequence. The acylation of benzylidene derivatives of sugars, such as **1** and **2** (Figure 2), useful intermediates in the synthesis of oligosaccharides, occurs with significant regioselectivity. For instance, as it has been described in Protocol 1, acylation of **1** with vinyl acetate by action of Pseudomonas cepacia lipase gave quantitatively the 2-*O*-acetate **1a**. Results obtained with similar sugar derivatives are reported in Table 2. These data were obtained independently by Riva and coworkers [15] and by Roberts and coworkers [20]. More recently, the same selective acylation at the C-3 hydroxyl group of ethyl 4,6-*O*-benzylidene-1-thio-β-D-glucopyranoside **3** [21] and phenyl 4,6-*O*-benzylidene-1-thio-β-D-glucopyranoside **4** [22] has also been reported.

Table 2   Lipase catalysed esterification of benzylidene glycopyranosides[a]

Substrate	% yield 2-O-Acetate	% yield 3-O-Acetate
Methyl 4,6-O-benzyliden-α-D-glucopyranoside	100	0
Methyl 4,6-O-benzyliden-β-D-glucopyranoside	6	94
Allyl 4,6-O-benzyliden-α-D-galactopyranoside	0	0
Allyl 4,6-O-benzyliden-β-D-galactopyranoside	2	98
Methyl 4,6-O-benzyliden-α-D-mannopyranoside	97	3
Methyl 4,6-O-benzyliden-β-D-mannopyranoside	2	98

See [14, 19].

In the large majority of the examples reported in the literature, the acylating agents are simple aliphatic acids, usually acetate and butanoate. However, hydrolases can catalyze transesterifications even with different acyl moieties, some of which are reported in Figure 3.

For instance, methyl 4,6-O-benzyliden-α-D-glucopyranoside (**1**), as well as its β analogue, have been acylated with different synthetically useful esters, such as benzoate (**9**), chloroacetate (**10**), pivaloate (**11**), and levulinate (**12**) [23]. Alkoxycarbonylation (introduction of Cbz) of 2-deoxysugars has been reported by Gotor, using his usual oxime derivatives (specifically, **13**) [24]. Selective acylation of sugars with amino-acids has been initially reported by Riva et al. [5], and more recently by Monsan and co-workers [25]. In the example described in Protocol 2, the use of two different lipases and two different activated esters allowed the preparation of the lactose derivative **15b**, selectively acylated at C-6′ OH and C-2′ OH with protective groups removable under different conditions, not necessarily requiring alkaline hydrolysis (Figure 4). As a general remark, it can be said that hydrolases show a high versatility towards the acylating agent. However, generally speaking, it is not true that it is possible to acylate any substrate with any kind of ester. As it has been pointed out in different reports [26], it is likely that a reciprocal steric hindrance occurs between large acyl groups and large nucleophiles, the latter being excluded by the catalytic site and thus prevented to attack the acyl-enzyme intermediate.

Protocol 3 gives an illustration of the complementary regioselectivity of different hydrolases on the same substrate. The benzyl amide of lactobionic acid **16** was selectively acylated to each of the two primary OH to give the monoesters **16a** and **16b** by switching from lipase PS to Novozym 435, while the protease subtilisin gave preferentially the diester **16c** (Figure 5).

Glycosides of various classes of natural products are widely distributed in nature, where they are often found present acylated with aliphatic and aromatic acids (mainly acetic, malonic, p-coumaric and ferulic) at specific hydroxyl groups of their sugar moieties [27]. Many of these compounds are bioactive molecules or possess other interesting properties. In nature, the formation of these esters is the last step in the biosynthetic pathway and it is catalyzed by different acyltransferases, enzymes that show relative flexibility towards the acyl group, but strict selectivity for the substrate to be esterified [28]. Moreover, they require stoichiometric amounts of the corresponding acyl-coenzyme A, and therefore they are not very convenient for *in vitro* laboratory synthesis. On the other hand, direct selective chemical acylation of glycosides is still a distant target because of the present lack of suitable reagents and protocols [29].

In one of the above mentioned papers [5], Riva et al. showed that subtilisin was also able to acylate natural glycosides in regioselective fashion. The reaction was performed on four model substrates: salicin, riboflavin, and the nucleosides adenosine and uridine (Figure 1). Not only subtilisin was reactive with these molecules, but it also showed absolute selectivity for their sugar moieties even in the presence of reactive functional groups on the aglycons (like in salicin and adenosine). The same protocol was later applied to other natural glycosides.

Protocol 4 and Protocol 5 describe the selective modification of two flavonoid glycosides, an important group of natural compounds widely distributed in the plant kingdom. In the first example (Protocol 4), the flavonoid disaccharide monoglycoside rutin (**17**) was selectively acylated by Novozym 435 to give the diacetate **17a** as the only product in more than 90% isolated yield. Only two of the six secondary and four aromatic hydroxyls of the molecule were selectively esterified by this lipase.

The second example (Protocol 5) presents a chemo-enzymatic approach to the synthesis of a specific natural compound, that is naringin 6''-*O*-malonate (**18b**). The two-step synthesis is based on the regioselective enzymatic introduction of a benzylmalonyl group to naringin (**18**) to give the ester **18a** (catalyzed by Novozym 435 lipase), followed by Pd/C hydrogenolysis of the benzyl moiety (Figure 7). This simple and high-yielding protocol for the selective malonylation has been applied to other natural glycosides, such as ginsenosides, a class of dammarane-type triterpene oligoglycosides which are isolated from *Panax ginseng C.A. Meyer* a plant widely used in the traditional Chinese medicine [30].

4.2 Glycosidases-catalyzed synthesis of alkyl glycopyranosides

Protocol 6 and Protocol 7 describe two examples of the use of glycosidases for the preparation of glycopyranosides. By using the alcohol donor in a large excess, it is possible to obtain the corresponding glycoside in just one step (the usual chemical protocols would require three steps at least) and with total stereocontrol of the anomeric bond formation.

Allyl, benzyl and trimethylsilylethyl glycopyranosides are important starting intermediates in carbohydrate chemistry as temporary anomeric protected derivatives. Specifically, Protocol 6 describes the facile and high-yielding preparation of allyl β-ᴅ-glucopyranoside (**19**, Figure 8) following the procedure suggested by Vic and Crout [13].

The same methodology has been applied for the preparation of *N*-(acryloylaminoethoxy)ethyl-β-ᴅ-glucopyranoside (**20**, Figure 8), a compound that has been used at IBRM for the preparation of a new polyacrylamide-based matrix for the electrophoretic separation of DNA fragments [17]. In this case the product **20** was isolated in low yields (only 18%), but the enzymatic process was still more convenient than the usual chemical protocols.

In conclusion, the selected examples described in this chapter show that the use of hydrolases in organic solvents for the selective acylation of sugars and of sugar derivatives as well as the use of glycosidases for the preparation of alkyl glycopyranosides are effective synthetic tools, these enzymes deserving a place in the lab along with the classical chemical reagents for the selective modification of carbohydrates.

Acknowledgments

We thank the Agriculture and Fisheries Program of the European Commission for financial support to part of the experimental work described in this chapter (FAIR CT 96–1048).

References

1   Wong C-H, Halcomb RL, Ichikawa Y, Kajimoto T (1995) Enzymes in organic synthesis: Application to the problems of carbohydrate recognition (Part 1). *Angew Chem Int Ed Engl* 34: 412–432

2   Nishikowa Y, Yoshimoto K, Ohkawa M (1979) Chemical and biochemical studies on carbohydrate esters. VII. Plant growth inhibition by an anomeric mixture of synthetic 1-*O*-lauryl-ᴅ-glucose. *Chem Pharm Bull* 27: 2011–2015

3   Therisod M, Klibanov AM (1986) Facile enzymatic preparation of monoacylated sugars in pyridine. *J Am Chem Soc* 108: 5638–5640

4   Therisod M, Klibanov AM (1986) Regioselective acylation of secondary hydroxyl groups in sugars catalyzed by lipases in organic solvents. *J Am Chem Soc* 109: 3977–3981

5   Riva S, Chopineau J, Kieboom APG, Klibanov AM (1988) Protease-catalyzed regioselective esterification of sugars and related compounds in anhydrous dimethylformamide. *J Am Chem Soc* 110: 584–589

6   Riva S (1995) Regioselectivity of hydrolases in organic media. In: AMP Koskinen, AM Klibanov (eds): *Enzymatic reactions in organic media*. Blackie A & P, London, 140–169

7   Rich JO, Bedell BA, Dordick JS (1995) Controlling enzyme-catalyzed regioselectivity in sugar ester synthesis. *Biotechnol Bioeng* 45: 426–434

8   Cai S, Hakomori S, Toyokuni T (1992) Application of protease-catalyzed esterification in synthesis of 6'-deoxy-6'-fluoro and 6-deoxy-6-fluorolactosides. *J Org Chem* 57: 3431–3437

9   Danieli B, Bertario A, Carrea G, Riva S (1993) Chemo-enzymatic synthesis of 6"-*O*-(3-arylprop-2-enoyl) derivatives of the flavonol glucoside isoquercitrin. *Helv Chim Acta* 76: 2981–2991

10 Adelhorst K, Bjorkling F, Godtfredsen SE, Kirk O (1990) Enzyme catalysed preparation of 6-*O*-acyl-glucopyranosides. *Synthesis* 112–115

11 Chen X, Johnson A, Dordick JS, Rethwisch DG (1994) Chemoenzymatic synthesis of poly(sucrose acrylate): optimization of enzyme activity and polymerization conditions. *Macromol Chem Phys* 195: 3567–3578

12 Fernandez-Mayoralas A (1997) Synthesis and modification of carbohydrates using glycosidases and lipases. *Topics Current Chem* 186 : 1–20

13 Vic G, Crout DHG (1995) Synthesis of allyl and benzyl β-D-glucopyranosides, and allyl β-D-galactopyranoside from D-glucose or D-galactose and the corresponding alcohol using almond β-D-glucosidase. *Carbohydr Res* 279: 315–319

14 Nilsson KGI (1988) A simple strategy for changing the regioselectivity of glycosidase-catalysed formation of disaccharides: part II, enzymatic synthesis in situ of various acceptor glycosides. *Carbohydr Res* 180: 53–59

15 Panza L, Luisetti M, Crociati E, Riva S (1993) Selective acylation of 4,6-*O*-benzylidene glycopyranosides by enzymatic catalysis. *J Carbohydr Chem* 12: 125–130

16 Lay L, Panza L, Riva S et al. (1996) Regioselective acylation of disaccharides by enzymatic transesterification. *Carbohydr Res* 291: 197–204

17 Chiari M, Riva S, Gelain A et al. (1997) Separations of DNA fragments by capillary electrophoresis in *N*-substituted polyacrylamides. *J Chromatogr A* 781: 347–355

18 Danieli B, Luisetti M, Sampognaro G et al. (1997) Regioselective acylation of polyhydroxylated natural compounds catalyzed by *Candida antarctica* lipase B (Novozym 435) in organic solvents. *J Mol Catalysis B: Enzymatic* 3: 193–201

19 Riva S, Danieli B, Luisetti M (1996) A two-step efficient chemoenzymatic synthesis of flavonoid glycoside malonates. *J Nat Prod* 59: 618–621

20 Iacazio G, Roberts SM (1993) Investigation of the regioselectivity of some esterifications involving methyl 4,6-*O*-benzylidene-D-pyranosides and *Pseudomonas fluorescens* lipase. *J Chem Soc Perkin Trans I* 1099–1101

21 Matsuo I, Isomura M, Walton R, Ajisaka K (1996) A new strategy for the synthesis of the core trisaccharide of asparagine-linked sugar chains. *Tetrahedron Lett* 37: 8795–8798

22 Gridley JJ, Hacking AJ, Osborn HMI, Spackman DG (1997) Regioselective lipase-catalysed acylation of 4,6-*O*-benzylidene-β-D-pyranoside derivatives displaying a range of anomeric substituents. *Synlett* 1397–1399

23 Panza L, Brasca S, Riva S, Russo G (1993) Selective lipase-catalyzed acylation of 4,6-*O*-benzylidene-β-D-glucopyranosides to synthetically useful esters. *Tetrahedron Asymmetry* 4: 931–932

24 Pulido R, Gotor V (1994) Towards the selective acylation of secondary hydroxyl groups of carbohydrates using oxime esters in an enzyme-catalyzed process. *Carbohydr Res* 252 55–68

25 Fabre J, Paul F, Monsan P, Perie J (1994) Enzymatic synthesis of amino acid ester of butyl α-D-glucopyranoside. *Tetrahedron Lett* 35: 3535–3536

26 Danieli B, Lesma G, Luisetti M, Riva S (1997) *Candida antarctica* lipase B catalyses the regioselective esterification of ecdysteroids at the C-2 OH. *Tetrahedron* 53: 5855–5862

27 Hostettmann K, Marston A (eds) (1995): *Chemistry and pharmacology of natural products: saponins,* Cambridge University Press, Cambridge (UK)

28 Koester J, Bussmann R, Barz W (1984) Malonyl-coenzyme A: isoflavone 7-*O*-glucoside-6″-*O*-malonyltransferase from roots of check pea. *Arch Biochem Biophys* 234: 513–521

29 Vermes B, Chari V.M., Wagner H (1981) Structure elucidation and synthesis of flavonol acylglycosides. III). The synthesis of tiliroside. *Helv Chim Acta* 64: 1964–1967

30 Danieli B, Luisetti M, Riva S et al. (1995) Regioselective enzyme-mediated acylation of polyhydroxy natural compounds. A remarkable, highly efficient preparation of 6′-*O*-acetyl and 6′-*O*-carboxyacetyl ginsenoside Rg$_1$. *J Org Chem* 60: 3637–3642

10 Reversed Micelles as Microreactors: N-terminal Acylation of RNase A and its Characterization

Joël Chopineau, Bernard Lagoutte, Daniel Thomas
and Dominique Domurado

Contents

1 Introduction

Reversed micellar systems are composed of tiny aqueous droplets dispersed in organic solvent and stabilized by a monolayer of surfactant molecules. It is well established that enzymes and other biomolecules can be encapsulated inside the internal water compartment of reversed micelles [1–3]. Reversed micellar solutions are isotropic and optically transparent. The great interests of these supramolecular aggregates reside in their size in the range of those of proteins, and in our capacity to adjust their diameter within a 1-nm accuracy by adjusting the degree of hydration ($W_0 = [H_2O]/[surfactant]$) [4]. Since reversed micelles provide a controllable aqueous microenvironment dispersed inside a bulk organic phase, tailor-modification of protein molecules is possible.

Different problems of protein chemistry have been studied in these systems:

Methods and Tools in Biosciences and Medicine
Methods in non-aqueous enzymology, ed. by M. N. Gupta
© 2000 Birkhäuser Verlag Basel/Switzerland

- the refolding yields of RNases A and T1 have been improved by using reversed micelles, because isolation of denatured protein molecules from each other results in reduction of the intermolecular interactions, which often lead to aggregation in standard aqueous solutions [5, 6].

- several oligomeric enzymes dissociate into monomers or dimers during their entrapment into reversed micelles, allowing studies of individual subunits [7–10].

- reversed micelles are also an efficient medium to study individual glycosylated protein molecules [11].

- proteins were chemically modified in reversed micelles: attachment of organometallics to glucose oxidase [12], deamidation of triosephosphate isomerase [13].

Conjugating fatty acids, poorly soluble in water, to proteins, poorly soluble in organic solvents, is still a challenging task. Protein acylation can be undertaken using either classical chemical methods [14–17] or reversed micelles as microreactors [18, 19]. In micellar media, derivatization of proteins by water-insoluble reagents can be limited to a small number of sites. On the other hand, carrying out this reaction in water is difficult and leads to a much more extensively modified protein. As yet, reversed micelles were used to acylate enzymes [19–22], antibodies [23] and Fab fragments [24]. However, the resulting products were poorly characterized even in studies concerning RNase A [25].

We proposed a method based on the results obtained for RNase A myristoylation performed in the micellar system AOT/isooctane/borate-sodium hydroxyde buffer [19]. RNase A is a convenient model for a complete study of in vitro acylation. Its primary sequence has been elucidated long ago [26]: this enzyme is composed of 124 amino acid residues forming a single-chain polypeptide and it contains 10 ε-amino groups. Its molar mass calculated from its amino acid composition is 13,682 daltons. Its crystal structure was resolved [27] and it still serves as a model for protein crystallization [28]. Its catalytic activity [29] and renaturing capacity [5] were studied in AOT reversed micelles. It is not known to exhibit hydrophobic interactions with membranes naturally.

The first aim of our work was to study the interaction of an acylated protein with lipid bilayers. We wished to examine how different are the behaviors of native proteins and of their acylated counterparts, what specific role is played by the peptide backbone in these interactions, what modulation can be introduced by different aliphatic acyl residues (length, insaturation, geometry) in the interaction. The interaction between acylated derivatives and lipid bilayers is usually studied with peptides, not proteins, thus confirming the difficulty to obtain these products [30, 31].

The second aim of our work was to examine the possibility that the interaction of acylated proteins with lipid bilayers could lead to the crossing of cell membrane, possibly also to the crossing of physiological barriers such as the blood-brain barrier [32]. This would be of huge interest for enzyme therapy of the numerous genetic diseases affecting brain function.

Finally, it must be mentioned that protein acylation is a way to modulate the immunological response (loss of immunogenicity while antigenicity is conserved when short aliphatic chains are used [33]).

2   Materials

Chemicals

Acetone, HPLC grade	Riedel de Haen
Acetonitrile, HPLC grade	Riedel de Haen
AOT (sulfosuccinic acid bis [2-ethylhexyl] ester)	Sigma
Cytidine 2'-monophosphate	Sigma
Cytidine 2':3'-monophosphate	Sigma
Fatty acid chlorides (capryl, lauryl, myristoyl, palmitoyl, stearoyl, palmitoleyl)	Sigma
Trifluoroacetic acid, sequencing grade	Fluka
2,2,4-trimethyl pentane (isooctane) (dried on 3 Å molecular sieves)	Aldrich

Enzymes
- RNase A (EC 3.1.27.5) from bovine pancreas was purchased from Boehringer (Mannheim, Germany). RNase A concentration was routinely determined using a molar extinction coefficient of 9700 $M^{-1}cm^{-1}$ at 278 nm.

Equipment
- Absorbances were measured using an UVIKON 930 spectrophotometer (Kontron, Rotkeuz, Switzerland).

Solutions
- Deionized water was obtained from a MilliQ apparatus (Millipore, Milford, MA, USA).
- Borate-sodium hydroxide buffers (50 mM, 7.8 < pH < 10.1) were used for protein incorporation in reversed micelles.

All other reagents were of analytical grade.

3 Methods

3.1 Protocols

Protocol 1   RNase A modification

The reactions were carried out in 50-ml polytetrafluoethylene centrifugation tubes. RNase A was incorporated into AOT reversed micelles using the injection method [1–3]. In this way, different volumes of the protein solution at different molarities in 50 mM borate-sodium hydroxide buffer were solubilized in 25 ml of 0.1 M solution of AOT in isooctane. Then 1 ml of isooctane saturated with fatty acid chloride was added. Different acylation-reagent: protein molar ratios were selected. The solution obtained was stirred during 2 hrs on a rocking-stirrer at room temperature. The protein was precipitated from the reaction mixture by addition of 25 mL of cold acetone (-20 °C). The protein precipitate was collected after a 20-min centrifugation at 2000 g and -20 °C, then washed five times as follows: addition of 10 mL of cold acetone (-20 °C), 10 min of centrifugation at 2000 g and -20 °C, the supernatant containing isooctane and AOT being discarded. AOT presence in the different fractions was checked by thin layer chromatography. Residual acetone was evaporated at room temperature under vacuum.

Protocol 2   RP-HPLC analysis and purification of derivatives

Both protein analysis and purification were carried out using RP-HPLC. The column (Si C18 Nucleosil, 4.6 mm internal diameter, 25 cm length, 10 µm particle size, 30 nm porosity) was maintained at 30 °C. Protein elution was monitored at 280 nm.

For analysis the flow rate was 1 mL/min. Elution was obtained with a linear gradient: 0–2 min 100% A; 2–20 min 0% to 95% B, with A:0.05% of trifluoroacetic acid in water and B:0.05% of trifluoroacetic acid in acetonitrile.

The same device was used for protein purification. A 100-µL loop was loaded with 4 mg of freeze-dried protein dissolved in water. Samples were eluted at 3 ml/min with a four-step isocratic gradient (25% B 10 min, 30% B 14 min, 35% B 30 min, 100% B 15 min). The two major fractions were collected separately. Acetonitrile was evaporated under a nitrogen stream and water was removed by freeze-drying.

Protocol 3   Characterization of acylated derivatives

Electrophoreses

SDS polyacrylamide gel electrophoresis of modified RNase A samples were performed on a high density phastgel® using a PhastSystem (Pharmacia, Sweden). Molar mass markers were in the range 2,512–16,949 daltons.

RNase A derivatives (1 mg/ml) were analysed by capillary electrophoresis using a Beckman PACE instrument under the following conditions: fused silica capillary (0.37 m x 75 μm internal diameter), buffer was sodium phosphate 20 mM, sodium borate 5 mM pH 7.2; operating voltages: 5 kV for 10 sec and 7 kV for 30 min, temperature was 30 °C and detection was performed at 214 nm.

Electro-spray ionisation mass spectrometry (ESI-MS)

Each sample was dissolved in methanol/water – 1% acetic acid (50/50 v/v) and injected in a Hewlett-Packard Engine 5989 mass spectrometer (Palo Alto, CA, USA) equipped with an electrospray inlet system (Analytica of Brandford). The spectra were recorded in the 1000–1900 range of mass-to-charge (m/z) ratios in steps of 0.10 m/z, with a 2 ms dwell time and with a resolution of 1 mass unit. The instrument was tuned and calibrated in the positive mode (for the detection of cations). Samples were infused into the source using a Harvard 22 syringe pump (South Natick, MA, USA) at a flow rate of 2 μL/min.

Free amino groups titration

Free amino groups of modified RNase A were assessed using a protocol [19] adapted from Fields [34].

Peptide mapping and N-terminal sequencing

For peptide mapping, disulfide bridges of sample proteins were first reduced overnight and the free cysteines were blocked with vinyl pyridine. Cleavage after glutamates was performed overnight at 37 °C in 50 mM ammonium carbonate buffer pH 8.0 [35]. Peptides were separated on a C18 Delta Pack column (Waters) using a 40 min 0–64% acetonitrile linear gradient in H$_2$O/0.1% trifluoroacetic acid at an elution rate of 1 mL/min. Edman procedure was performed automatically on 2 nmoles samples with a pulsed liquid sequencer (Applied Biosystems, Roissy Charles de Gaulle, France). Phenylthiohydantoin amino acids were separated on line by RP-HPLC and were quantified at 256 nm [36].

Enzymatic activity determination

Enzyme activity of native and acylated RNase A samples was determined using cytidine 2':3'-monophosphate as a substrate. Initial velocity was calculated from the changes in absorbance at 292 nm, using an absorbance difference of 570 M^{-1}cm^{-1} between cytidine 2':3'-monophosphate and cytidine 3'-monophosphate at this wavelength [37]. Substrate concentrations were controlled using a molar extinction coefficient of 8,400 M^{-1}cm^{-1} at 268 nm [38]. Initial velocity measurements were done in triplicate using 27.5 μg of enzyme in 1 mL Tris buffer (0.1 M, pH 7), for a substrate concentration ranging from 0.1 to 1 mM. The kinetic constants V_{max} and K_M were determined by non linear regression [39].

4   Remarks and Conclusions

In this chapter, we report the controlled chemical modification of RNase A. Using reversed micelles as aqueous microreactors, the acylation process comprised three major steps:

1) The protein was incorporated into reversed-micelles by the injection method, and an organic solution of the acyl chloride, the acylating agent, was added.
2) Through precipitation using cold acetone, the modified protein was recovered from the reaction mixture as a hydrosoluble fraction.
3) The hydrosoluble acylated derivatives of RNase A were purified using semi-preparative RP-HPLC.

These three steps were optimized in order to obtain high purity acylated RNase A derivatives. This modification was monitored using RP-HPLC as an analytical method.

4.1   Optimization of protein modification and recovery

The first two steps are linked and were examined together. W_0, the water/surfactant ratio, is a key parameter when tailoring proteins in micellar media [5–13, 18–25] because the size of the inner aqueous compartment of reversed micelles depends on it. Influence of this parameter for enzymatic catalysis in microheterogeneous systems has been demonstrated for more than 40 enzymes [1–3]. Consequently, dependence on W_0 of protein recovery and modification degree has been studied.

The yield of native RNase A recovery from reversed micelles through acetone precipitation was examined for W_0 values between 6 and 25. At $W_0 = 6$ incorporation of native RNase A into reversed micelles is possible but acetone failed to precipitate the protein from the organic solvent/surfactant mixture. For higher W_0 values, recovery (70 to 85%) did not depend on W_0. Enzymatic activity was not affected by acetone precipitation.

In the same 6–25 W_0 range, RNase A myristoylation increased when W_0 decreased with an optimum for $W_0 = 7$. The lowest $W_0 (= 7)$, which allowed us to recover the protein from AOT reversed micelles, was then found to be optimum for both acylation and recovery as a soluble fraction [19]. This W_0 value corresponds to a theoretical inner radius of empty reversed micelles of 14.5 Å [16], whereas the gyration radius of RNase A is 18 Å. Incorporation of RNase A into the aqueous compartment of the micellar medium probably resulted in the formation of larger micelles as noticed by pioneers of the field [29].

Protein content is an important parameter for tailoring macromolecules in reversed-micelles microreactors as described for RNase A and T1 refolding [5, 6] or for triosephosphate isomerase reactivation [9]. At $W_0 = 7$, 10 mg of protein in 25 ml of isooctane was determined to be the optimum for RNase A myristoylation. In these conditions, more than 60% of modified protein was recovered using a 4:1 myristoyl chloride/RNase A molar ratio.

Quantities of recovered native and modified protein are function of the molar reagent/protein ratio. As this ratio increased, the number of modified sites on the protein increased, this leading to a decrease in the water soluble fraction. Modification degree and protein recovery are correlated. The yield of modified protein recovered is decreasing when the reagent ratio is higher than 4:1 even if the degree of modification is enhanced. Protein modification by acylating agents are usually performed using a large excess of reagent in order to get a high modification degree, for example in the case of L-asparaginase acylation with palmitoyl chloride a 200-fold excess was used [14].

4.2 Acylated derivatives purification and characterization

The protein recovered by acetone precipitation contained two major products, the unreacted native RNase A and its acylated counterpart, which were separated using RP-HPLC. Capryl, lauryl, myristoyl, palmitoyl, stearoyl and palmitoleyl modifications allowed us to produce acylated RNases A of different hydrophobicity. After freeze-drying, the purity of each hydrophobic derivative was controlled by analytical RP-HPLC and by capillary electrophoresis (Fig. 1). In order to identify the modified protein, two standards were used, namely the commercial native RNase A and the RNase A having undergone the modification process without acylating reagent.

The commercial native protein batch contained more than 99% of RNase A as demonstrated by both methods. RP-HPLC-purified capryl, lauryl, myristoyl, palmitoyl, stearoyl and palmitoleyl derivatives of RNase A contained less than 1% of unreacted protein as shown by RP-HPLC (Figure 1A for myristoyl derivative). When using RP-HPLC, a difference in hydrophobicity among the derivatives could be observed [19, 20].

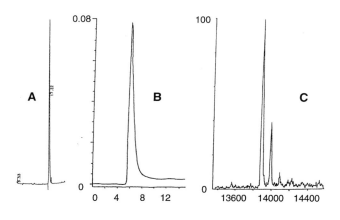

Figure 1 Analysis of myristoylated RNase A

The myristoyl derivative purity is checked by RP-HPLC (A), by Capillary Electrophoresis (B) and Electro-Spray Ionization Mass Spectrometry (C).

In capillary electrophoresis, the presence of fatty acid moiety(ies) diminishes the retention time (5.8 min against 6.3 min) of the modified proteins when compared to the standards. However the retention time was identical for the different hydrophobized derivatives (Figure 1B for myristoyl derivative, data not shown).

From post reaction amino groups titration, we concluded that hydrophobic derivatives of RNase A bear 1–2 fatty acid residues per protein molecule [19]. In order to determine exactly this ratio, ESI-MS was performed on each purified derivative. This technique is known for its accuracy and the identification of possible covalent modifications [40].

ESI-MS spectrum for myristoylated RNase A is shown in Figure 1C and mass data for all derivatives are collected in the second column of Table 1. The heterogeneous nature of the mass spectrum is consistent with the presence of 2–4 molecular species containing respectively 0, 1, 2 or 3 phosphate or sulfate group(s) non covalently bound to the protein as previously shown by others [41]. These anions are most probably strongly bound to the basic lysine or arginine residues of the protein, particularly in the active site. They are partly removed during the modification process as attested by the results concerning control RNase A and all derivatives: these preparations carry a maximum of one anion per protein molecule.

Table 1 Comparison of native, standard and acylated RNases A in term of molar masses

Measured molar masses, mass variation (after subtraction of the molar mass of native RNase A = 13,682 Da) and expected mass variation (due to the presence of 0–3 phosphate or sulfate groups bound to the native enzyme and to the presence of fatty acid moieties).

Putative sample	Molar mass (Da)	Mass variation (Da)	Expected mass variation (Da)	Phosphate (or sulphate) groups	Fatty acid moieties
Native	13684 ± 2	2	0	0	0
RNase A	13781 ± 2	99	97	1	0
	13878 ± 1	196	194	2	0
	13977 ± 4	295	291	3	0
Control	13682 ± 1	0	0	0	0
RNase A	13779 ± 1	97	97	1	0
Caprylated	13836 ± 1	154	154	0	1
RNase A	13931 ± 1	249	251	1	1
Laurylated	13864 ± 1	182	182	0	1
RNase A	13959 ± 1	277	279	1	1
Myristoylated	13897 ± 1	215	210	0	1
RNase A	13997 ± 1	315	307	1	1
Palmitoleylated	13918 ± 2	236	236	0	1
RNase A	14016 ± 1	334	333	1	1
Palmitoylated	13920 ± 2	238	238	0	1
RNase A	14018 ± 1	336	335	1	1
Stearoylated	13948 ± 2	266	266	0	1
RNase A	14045 ± 1	363	363	1	1

After subtraction of the anion mass from the total molar mass of each derivative, we are left with a residual variation (column 3 of Table 1). A mass variation of 154, 182, 210, 238, 266 and 236 was respectively expected for the fixation of a single capric, lauric, myristic, palmitic, stearic and palmitoleic fatty acid moiety per protein molecule [42]. The mass data for all samples are consistent with the presence of a single fatty acid residue per protein molecule.

The mass data cannot rule out the possibility of mixtures of monosubstituted polypeptide chains and do not give the position(s) of the substitution(s). Sequencing experiments were then performed to address these two important questions.

In order to determine the exact location of the linked acyl chain, we compared the peptide maps of control and of palmitoylated RNases A, both cleaved preferentially on the carboxylic side of glutamyl residues by protease from *Staphylococcus aureus* V8 [35]. The only difference between both maps was the complete absence of the *N*-terminal nonapeptide (KETAAAKFE) at the normal elution time (Fig. 2A), strongly suggesting a modification of either lysine 1 or lysine 7. Application of automatic Edman degradation to control and acylated RNases A resulted in a very low yield of lysine 1 recovery for all acylated samples never higher than 2.4% of the control (Fig. 2B). This is likely to be the consequence of a covalent modification of the *N*-terminal α-amino group, the only modification susceptible to impair Edman degradation, in contrast to ε-amino group modifications. This full set of results strongly supports the single and specific homogeneous acylation of the α-amino group of RNase A lysine 1.

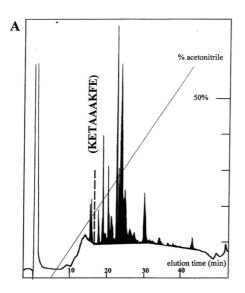

A

Figure 2 Analysis of control and acylated Rnases A using peptide mapping and N-terminal sequencing.

(A) Peptide map of palmitoylated RNase A cleaved after glutamate residues by the protease from *Staphylococcus aureus* V8. The dotted line show the expected place of the unmodified *N*-terminal nonapeptide, as observed on a control sample.
(B) Comparative sequencing of control, myristoylated, palmiltoylated and stearoylated derivatives of RNase A (C14, C16 and C18 - Rnase, respectively) performed on 2 nmoles of protein. The very low yield of Lys 1 recovery in all acylated samples did not allow an accurate determination of the following residues.

B

N-terminal sequencing of native and acylated RNases (pm of PTH aa)				
sample PTH	lys 1	glu 2	thr 3	ala 4
control RNase	675	502	403	549
C14 - RNase	10	-	-	-
C16 - RNase	16	-	-	-
C18 - RNase	11	-	-	-

Indeed, eleven potential sites of acylation are present on RNase A: 10 ε-amino groups and the α-amino group on the *N*-terminal lysine. From the RNase A crystal structure, it is apparent that the most accessible amino acid of the protein is lysine 1, a good candidate for the reaction with acyl chlorides. Both α and ε-amino groups of this lysine could be the nucleophilic partner of this acylation, but their reactivities are expected to be different due to well-known differences in their respective pKs [43]. This is particularly true for RNase A. Its α-amino group was previously shown to react with fluorescein-isothiocyanate at pH 6.0 [44]. The optimal acylation pH in reversed micelles was obtained using an aqueous phase buffered at pH 10.1. However, previous results concerning RNase A in the same reversed micellar system demonstrated that the local pH inside the reversed micelles was lower than in the original aqueous phase by 2 to 3 pH units [29]. Moreover, the initial pH decreased as the acylation reaction takes place, due to the release of hydrogen chloride. In fact, the mean local pH was probably just compatible with the presence of the nucleophilic unprotonated form of the α-amino group. In these conditions, we could expect preferential α-amino acylations. Stable substitution on the α-amino group of a protein impairs the normal degradation by the Edman procedure, what has been observed indicative of a single fixation of the acyl chain on the *N*-terminal lysine of RNase A [20, 32].

Using cytidine 2':3'-monophosphate as substrate, all acylated preparations were found to be active and to follow a classical Michaelis-Menten mechanism. Kinetic parameters for each protein are reported in Table 2. The K_M was outside the range of substrate concentration for most samples, except for the palmitoleylated RNase A. This situation was not optimal for the identification of the kinetic parameters, but it was not possible to avoid this drawback consecutive to substrate solubility limitations. The control RNase A also exhibited a higher activity than the starting material. This might be related to the loss of phosphate (sulfate) anions during the extraction from the micelles.

Table 2   Steady-state kinetic parameters for hydrolysis of cytidine 2':3'-monophosphate by native, standard and monoacylated RNases A.

Sample	k_{cat} (second^{-1})	K_M (mM)	k_{cat}/K_M (second^{-1}.M^{-1})
Native RNase A	2.06 ± 0.11	2.03 ± 0.13	1010 ± 110
Control RNase A	1.79 ± 0.01	1.16 ± 0.01	1540 ± 25
Myristoylated RNase A	1.32 ± 0.08	2.10 ± 0.17	630 ± 80
Palmitoleylated RNase A	0.28 ± 0.03	0.29 ± 0.10	1000 ± 400
Palmitoylated RNase A	0.72 ± 0.02	1.5 ± 0.07	480 ± 40
Stearoylated RNase A	1.49 ± 0.01	1.47 ± 0.02	1010 ± 25

4.3   Conclusions

When choosing reversed micelles as microreactors, our major objective was to specifically achieve monoacylation of RNase A. In this system, the apolar solvent allows the dissolution of acyl chlorides of different lengths whereas the protein is solubilized in a dispersed microaqueous phase, resulting in a good accessibility of the protein to the lipophilic reagent. Also this procedure allows a fine tuning of the microreactor size whose diameter can be adjusted within a 1-nm accuracy by controlling the hydration degree (W_0 = [H$_2$O]/[surfactant]). Using optimal conditions [19], we produced pure RNase A derivatives. The chemical and crystalline structures of these new compounds have been extensively characterized together with their enzymatic activities [20, 32].

The very high resolution of capillary electrophoresis has shown the homogeneity of all fatty acid derivatives of RNase A, and their masses have been determined using ESI-MS. The mass data (Table 1) were all consistent with the presence of a single fatty acid residue per protein molecule, confirming the homogeneity observed by capillary electrophoresis. The sequencing results strongly argued for a specific acylation of the α-amino group of lysine 1. A single acyl chain was exclusively bound, with a high yield, to the N-terminal amino group of the polypeptide. Acyl RNases A were all found to be active towards cytidine 2':3'-monophosphate, demonstrating the usefulness of the method for the production of acyl derivatives with a preserved activity. These data pinpoint the high accuracy and efficiency of the overall process using reversed micelles as microreactors for producing monoacylated N-terminal RNase A derivatives.

The single acylation site observed in this work is partly the result of the lucky choice of RNase A, its N-terminal sequence being the most accessible of the protein. It also rests on the particular reactivity of the N-terminal amino group as compared to ϵ-amino groups. It will be interesting to check if this specific acylation procedure could become of general use for proteins with an accessible N-terminal, provided the local pH is correctly adjusted. Oncoming studies of several other proteins with a free N-terminal end will possibly validate this assumption. The precise spatial localization of the aliphatic chain was also investigated on crystals of the different derivatives [20]. This structural information is of the utmost importance for a precise understanding of membrane interactions of these new compounds. Among the numerous fields of interest, one of the most exciting is indeed the interaction of soluble acylated proteins with membranes, and the possible crossing of these fundamental biological barriers. This potential property of acylated proteins could be of great interest for therapeutic applications [32, 45].

Acknowledgments

The authors acknowledge the technical help of M.M. Flament and B. Becchi (Lyon) for mass spectrometry, of M. Vijayalakshmi and F. Roy (Compiègne) for capillary electrophoresis. We are also grateful to A. Menez for the access to the protein sequencer (Saclay).

Abbreviations

AOT	aerosol OT (sulfosuccinic acid bis [2-ethylhexyl] ester)
ESI-MS	electro-spray ionization mass spectrometry
RNase A	bovine pancreatic ribonuclease A (EC 3.1.27.5)
RP-HPLC	reversed phase-high performance liquid chromatography

References

1 Martinek K, Levashov AV, Klyachko Nl et al. (1986) Micellar enzymology. *Eur J Biochem* 155: 453–468

2 Luisi PL (1985) Macromolecules hosted in reverse micelles. *Angew Chem Int Ed Engl* 24: 439–450

3 Klyachko NL, Levashov AV, Kabanov AV et al. (1991) In: K Gratzel, K Kalyanasundaram (eds): *Kinetics and Catalysis in Microheterogeneous Systems*. Marcel Dekker, New-York. 135–181

4 Pileni MP (1993) Reverse micelles as microreactors. *J Phys Chem* 97: 6961–6973

5 Hagen AJ, Hatton TA, Wang DIC (1990) Protein refolding in reversed micelles. *Biotech Bioeng* 35: 955–965

6 Shastry MCR, Eftink MR (1996) Reversible thermal unfolding of ribonuclease T1 in reverse micelles. *Biochemistry* 35: 4094–4101

7 Kabanov AV, Nametkin SN, Evtushenko et al. (1989) A new strategy for the study of oligomeric enzymes: α-glutamyltransferase in reversed micelles of surfactants in organic solvents. *Biochim Biophys Acta* 996: 147–152

8 Kabakov VE, Merker S, Klyachko NL et al. (1992) Regulation of the supramolecular structure and the catalytic activity of penicillin acylase from *Echerichia coli* in the system of reversed micelles of Aerosol OT in octane. *FEBS Lett* 311: 209–212

9 Garza-Ramos G, Tuena de Gomez-Puyou M, Gomez-Puyou A, Gracy RW (1992) Dimerization and reactivation of triose phosphate isomerase in reverse micelles. *Eur J Biochem* 208: 389–395

10 Chang GG, Huang TM, Huang SM, Chou WY (1994) Dissociation of pigeon-liver malic enzymes in reverse micelles. *Eur J Biochem* 225: 1021–1027

11 Levashov AV, Rariy RV, Martinek K, Klyachko NL (1993) Artificially glycosylated α-chymotrypsin in reversed micelles of Aerosol OT in octane. *FEBS Lett* 336: 385–388

12 Ryabov AD, Trushkin AM, Baksheeva et al. (1992) Chemical attachement of organometallics to proteins in reversed micelles. *Angew Chem Int Ed Engl* 31: 789–790

13 Garza-Ramos G, Tuena de Gomez-Puyou M, Gomez- et al. (1994) Deamidation of triosephosphate isomerase in reversed micelles: effects of water on catalysis and molecular wear and tear. *Biochemistry* 33: 6960–6965

14 Martins MB, Jorge JS, Cruz ME, (1990) Acylation of L-asparaginase with total retention of enzymatic activity. *Biochimie* 72: 671–675

15 Plou FJ, Ballesteros A (1994) Acylation of subtilisin with long fatty acyl residues affects its activity and thermostability in aqueous medium. *FEBS Lett* 339: 200–204

16 Heveker N, Bonnafé D, Ullmann A (1994) Chemical fatty acylation confers hemolytic and toxic activities to adenylate cyclase protoxin of *Bordetella pertussis*. *J Biol Chem* 269: 32844–32847

17 Ekrami HM, Kennedy AR, Shen W-C (1995) Water-soluble fatty acid derivatives as acylating agents for reversible lipidization of polypetides. *FEBS Lett* 371: 283–286

18 Levashov AV, Kabanov AV, Nametkin SN et al. (1984) Chemical modification of proteins (enzymes) with water insoluble reagents. *Dokl Akad Nauk SSSR* 284: 744–758 (in Russian); *Dan* (Engl. ed.) (1985) 284: 306–309

19 Robert S, Domurado D, Thomas D, Chopineau J (1993) Fatty acid acylation of RNase A using reversed micelles as microreactors. *Biochem Biophys Res Comm* 196: 447–454

20 Roy M-O, Uppenberg J, Robert S et al. (1996) Crystallisation of monoacylated proteins: influence of acyl chain length. *Eur Biophys J* 26: 155–162

21 Slepenev VI, Phalente L, Labrousse H et al. (1995) Fatty acid acylated peroxydase as a model for the study of interactions of hydophobically-modified proteins with mammalian cells. *Bioconjugate Chem* 6: 608–615

22 Pitré F, Regnault C, Pileni MP (1993) Structural study of AOT reverse micelles containing native and modified α-chymotrypsin. *Langmuir* 9: 2855–2860

23 Kabanov AV, Ovcharenko AV, Melik-Hubarov NS et al. (1989) Fatty acid acylated antibodies against virus suppress its reproduction in cells. *FEBS Lett* 250: 238–240

24 Chekhonin VP, Kabanov AV, Zhirkov YA, Morozov GV (1991) Fatty acid acylated Fab fragments of antibodies to neurospecific proteins as carriers for neuroleptic targeted delivery in brain. *FEBS Lett* 287: 149–152

25 Michel F, Pileni MP (1994) Synthesis of hydophobic ribonuclease by using reverse micelles. Structural study of AOT reverses micelles containing ribonuclease derivatives. *Langmuir* 10: 390–394

26 Hirs CHW, Moore S, Stein WH (1960) The sequence of the amino acid residues in performic acid-oxidized ribonuclease. *J Biol Chem* 235: 633–647

27 Richards FM, Wyckoff HW (1971) Bovine pancreatic ribonuclease A. In: Boyer (ed): *Enzymes*. Academic Press, New York, 4: 647–806

28 Fedorov AA, Joseph-McCathy D, Fedorov E et al. (1996) Ionic interactions in crystalline bovine pancreatic ribonuclease A. *Biochemistry* 35: 15962–15979

29 Wolf R, Luisi PL (1979) Micellar solubilization of enzymes in hydrocarbon solvents. Enzymatic activity and spectroscopic properties of ribonuclease in n-octane. *Biochem Biophys Res Comm* 89: 209–217

30 Peitzsch RM, McLaughlin S (1993) Binding of acylated peptides and fatty acids to phospholipids vesicles: pertinence to myristoylated proteins. *Biochemistry* 32: 10436–10443

31 Sankaram MB (1994) Membrane interactions of small *N*-myristoylated peptides: implications for membrane anchoring and protein-protein association. *Biophys J* 67: 105–112

32 Chopineau J, Robert S, Fénart L et al. (1997) Physicochemical characterization and in vitro interaction with brain capillary endothelial cells of artificially monoacylated ribonucleases A. *Lett Pept Sci* 4: 313–321

33 Shi Q, Domurado M, Domurado D (1995) Effect of protein chemical hydrophobization on anti-glucose oxidase immunoglobulin production in mouse. *Pharmacol Toxicol* 76: 278–285

34 Fields R (1971) The measurement of amino groups in proteins and peptides. *Biochem J* 124: 581–590

35 Houmard J, Drapreau GR (1972) Staphylococcal protease: A proteolytic enzyme specific of glutamoyl bonds *Proc Nat Acad Sci USA* 69 : 3506–3509

36 Stone KL, LoPresti MB, Crawford et al. (1989) Enzymatic digestion of proteins and HPLC peptide isolation. In: Matsudaira PT (eds): *A practical guide to protein and peptide purification for microsequencing.*, Academic Press, London 33–47

37 Crook EM, Mathias AP, Rabin BR (1960) Spectrophotometric assay of bovine pancreatic ribonuclease by the use of cytidine 2':3'-phosphate. *Biochem J* 74: 234–238

38 Sage HJ, Singer SJ (1962) Properties of bovine pancreatic ribonuclease in ethylene-glycol solution. *Biochemistry* 1: 305–317

39 Purves RD (1993) Anomalous parameter estimates in the one compartment model with first order adsorption. *J Pharm Pharmacol* 45: 934–936

40 Ashton DS, Beddell CR, Green BN, Oliver RWA (1994) Rapid validation of molecular structures of biological samples by electrospray-mass spectrometry. *FEBS Lett* 342: 1–6

41 Camilleri P, Haskins NJ, Rudd PM, Saunders MR (1993) Applications of electrospray mass spectrometry to studies on the structural properties of Ribonuclease A and Ribonuclease B. *Rapid Commun Mass Spec* 7: 332–335

42 Jacquinod M, Van Dorsselaer A (1992) Acquis et perspectives de l'electrospray en chimie et en biochimie. *Analusis* 20: 407–412

43 Lehninger AL (1989) *Principles of Biochemistry*: Worth publishers Inc., New-York, 95–120

44 Garel J-R (1976) pK changes of ionizable reporter groups as an index of conformational changes in proteins. *Eur J Biochem* 70: 179–189

45 Chopineau J, Robert S, Fénart L et al. (1998) Monoacylation of ribonuclease A enables its transport across an in vitro model of the blood-brain barrier. *J Control Rel*; 56: 231–237

11 Analysis in Non-Aqueous Milieu Using Thermistors

Kumaran Ramanathan, Birgitta Rees Jönsson and Bengt Danielsson

Contents

1   Introduction

Bioanalytical processes using biosensors have generated immense interest since the first elucidation of enzyme-based sensor for glucose [1]. Thereafter biosensors based on enzymes have received increasing attention [2, 3]. Enzyme-based biosensors unify high catalytic conversion rates and selectivity of an enzyme for analyte detection, and are evaluated on the basis of their storage stability, sensitivity, response time, operational stability and linearity. One major goal in biosensors has been to detect and quantify molecules in very complex matrices. In order to achieve this, much of the stress is laid on sensor fabrication that includes choice of the enzyme, technique of immobilisation, the nature and type of interaction between the enzyme and the solvent. Earlier

Methods and Tools in Biosciences and Medicine
Methods in non-aqueous enzymology, ed. by M. N. Gupta
© 2000 Birkhäuser Verlag Basel/Switzerland

it was generally assumed that enzymes function only in aqueous media, and hence very little thought was given to biosensor-based analysis in organic solvents. In the past decade, however, it has been demonstrated by several workers that enzymes can function effectively well in organic phase as well as in aqueous phase [4]. In this regard an early work [5] in organic phase biosensors followed by other investigations on enzyme electrodes [6] encouraged important research inputs on the role of organic solvents in biosensor-based analysis. In most cases the investigations on organic phase biosensor (OPB) are focused on the effect of the type of organic solvent on the catalytic efficiency of the sensor. Selectivity vís-a-vís substrates, solubility of the substrate and detectability of the analytes in the complex samples without pretreatment are other significant issues.

The performance of enzymes in non-aqueous solvents has made it feasible to explore a plethora of applications in biosensor technology. The reactions that are impossible in aqueous assays due to several constraints are now carried out in suitable non-aqueous solvents. Further, in organic solvents enzyme specificity may be altered [7] by manipulating the reaction conditions such as the degree of hydration, polarity, and hydrophobicity of the organic solvent. Several advantages of using organic phase enzyme biosensors are highlighted in some of the recent articles [8, 9] It must be reiterated that the main utility of OPB is the ability to assay compounds directly that are insoluble or partially soluble in water, but are easily soluble in organic solvents [10, 11]. As a large number of organic solvents are available, by proper selection of the solvents, any analyte which is a substrate for an enzyme reaction may be detected with a suitable OPB [12, 13]. Hence the inclusion of OPB has greatly improved the versatility of biosensors.

1.1   Fundamentals of calorimetric devices

Reaction of molecules with each other results in an exchange of several forms of energy. One well-known form of energy is heat. The advantages of measuring heat (calorimetry) were identified several decades before [14]. Almost all interactions, either physical, chemical or biological, involve exchange of heat. Specifically, enzyme based catalysis is associated with rather high enthalpy changes. Depending on the type of the catalytic reaction either a single or a combination of enzymes could be employed for producing detectable heat/thermometric signal.

In previous investigations the scope of calorimetry, especially in routine analysis, was constrained due to expensive instrumentation, coupled with a relatively slow response time and high costs. Several such calorimetric devices based on immobilized enzymes were introduced in the early 70's that combined the general detection principle of calorimetry with enzyme catalysis [15]. These instruments were useful due to the reusability of the biocatalyst, a continuous flow operation, inertness to optical and electrochemical interference and simple operating procedures. In the following years several of these concepts were further developed which culminated in the realisation of the ET, designed in our laboratory [16]. The technique drew immediate attention in the area of biosensors and has been successfully exploited for the past two decades [17]. The initial demonstration of the approach focused on the determination of glucose and urea. Subsequently, the ET has been applied for the determination of a wide variety of molecules [18].

More recently, several miniaturized prototypes of the ET have been fabricated e.g. a thermal probe designed as an integrated circuit, called a biocalorimetric sensor for glucose with total dimensions of only $1 \times 1 \times 0.3$ mm has been designed [19]. In another model a small thermoelectric sensor that employs a thin-film thermopile to measure the evolved heat was described [20]. These devices are less affected by extrinsic thermal effects compared to thermistor-based calorimetric sensors and could be operated without external temperature control. Active work is in progress in the authors' laboratory to construct a miniaturised portable bio-thermal flow injection system suitable for on-line monitoring. An instrument with 0.1–0.2 mm (ID) flow channels and a flow rate of 25–30 μL/min with sample volumes of 1–10 μL is being evaluated at present. A 1×5 mm enzyme column allows determination of glucose concentrations down to 0.1 mM. Recently a device equipped with thin-film temperature sensors of thermistor type (0.1×0.1 mm or smaller) for simultaneous glucose, urea and penicillin measurements has also been developed [21].

1.2   Principle of calorimetric measurements

In calorimetric assays the total heat evolved is proportional to the molar enthalpy as total number of product molecules created in the reaction.

$$Q = -n_p(\Delta H) \tag{1}$$

where Q = total heat, n_p = moles product, and ΔH = molar enthalpy change. It is also dependent on the heat capacity C_p of the system including the solvent:

$$Q = C_p (\Delta T) \tag{2}$$

The change in temperature (ΔT) recorded by the ET is thus proportional to the enthalpy change and on the heat capacity of the reaction.

$$\Delta T = - \Delta H \, n_p / C_p \tag{3}$$

As the heat capacity of most organic solvents is two or three times lower than that of water, an enhanced sensitivity is expected when using organic solvents [9]. This is the case, provided, the ΔH remains unaltered.

Table 1 Expected calorimetric sensitivities in various organic solvents.

Solvent	C_p (J mL^{-1}K^{-1})	Relative Sensitivity
Acetic acid	2.16	1.93
Acetone	1.70	2.46
Benzene	1.53	2.73
Carbon disulphide	1.62	2.58
Carbon tetrachloride	1.36	3.07
Chloroform	1.44	2.90
1,4-Dichlorobenzene	1.21	3.46
Diethyl ether	1.65	2.53
Ethanol	1.96	2.13
Ethyl acetate	1.74	2.40
n-Hexane	1.49	2.80
Methanol	2.02	2.07
Toluene	1.56	2.68
Water	4.18	1.00

Table 1 is a list of the heat capacity values of a few enzyme-catalysed reactions in organic solvents. A thermometric measurement is based on the sum of all enthalpy changes in the reaction mixture. Thus, it is useful to co-immobilise an oxidase with catalase, which results in doubling the sensitivity by replenishing part of the oxygen consumed in the reaction.

$$\text{β-D-glucose} + O_2 \quad \xrightarrow{\text{GOD}} \quad \text{β-D-gluconolactone} + H_2O_2 \tag{4}$$

$$2H_2O_2 \quad \xrightarrow{\text{Catalase}} \quad 2H_2O + O_2$$

The high protonation enthalpy of buffer ions like Tris could be utilised to enhance the total enthalpy of proton-producing reactions. A notable increase in the sensitivity could also be obtained in substrate or coenzyme recycling enzyme systems, in which the net enthalpy change of each turn in the cycle adds to the overall enthalpy change [22].

An inherent disadvantage of calorimetry is the lack of specificity. All enthalpy changes in the reaction mixture contribute to the final measurement. It is therefore essential to avoid non-specific enthalpy changes due to dilution or solvation effects. In most cases this is not a serious problem. An efficient way of coping with non-specific effects is to incorporate a reference column with an inactive filling and carrying out a differential determination.

The flow injection technique is usually employed for an ET assay. The sample volumes employed are too small to give any thermal steady state but generate a temperature peak [23]. The peak height of the thermometric recording is proportional to the enthalpy change corresponding to a specific substrate concentration. In most instances, the area under the peak and the ascending slope of the peak have also been found to be linearly varying with the substrate concentration. A sample introduction of sufficient duration (several minutes) leads to a thermal steady state resulting in a temperature change, proportional up to a certain substrate concentration.

1.3   Selection of specific solvents for thermal biosensing

Variation in solvent hydrophobicity, dielectric constant and water content of the solvent system may affect the ability of enzymes to use their free energy of binding with substrate. This may alter substrate specificity and reactivity. The solvent medium for OPB could be divided into; anhydrous organic media and water-containing organic media. The former includes measurements in pure solvents or mixture of pure organic solvents. These may be polar or non-polar in nature depending on the end application. The water-containing organic media comprise of water-organic solvents mixtures, water and an immiscible organic solvent forming a biphasic system or reverse micellar solutions. Enzymes generally need some water of hydration which is essential for their activity. Non-polar solvents are usually saturated with water or mixed with 1% water before use as reaction media for OPB. The water to organic phase ratio is very small, and it depends on the ability of the solvent to absorb water. OPB demonstrate greater reactivity in polar solvents in the presence of some amount of water [24–26]. It has been observed by various researchers, that hydrophobic solvents with dispersed water are better as biosensing media to maintain the active site environment of the enzyme. This further suggests that the dehydration effect of hydrophilic organic phase can be avoided by mixing adequate amount of water. Our studies on ET have also demonstrated such effects of hydrophilic organic solvents [7].

1.4   The transducer

Thermometry
The instrumentation for the ET fabrication normally employs a thermistor as a temperature transducer. Thermistors are resistors with very high negative temperature coefficient of resistance. These resistors are ceramic semiconductors made by sintering mixtures of metal oxides from manganese, nickel, cobalt, copper, iron and uranium. These could be obtained in many different configurations, sizes (down to 0.1–0.3 mm beads) and varying resistance values from such manufacturers. The best empirical expression to date describing the resistance-temperature relationship is the Steinhart-Hart equation [27]:

$$1/T = A + B(\ln R) + C (\ln R)^3 \tag{6}$$

Where T = temperature (K); ln R = the natural logarithm of the resistance, and A, B and C are derived coefficients. For narrow temperature ranges the above relationship could be approximated by the equation [27]:

$$R_T = R_{To} \, e^{\beta(1/T-1/T_o)} \tag{7}$$

where R_T and R_{To} are the zero-power resistances at the absolute temperatures T and T_0, respectively, and "β" is a material constant that ranges between 4000 and 5000 K for most thermistor materials. This yields a temperature coefficient of resistance between −3 and −5.7% per Celsius degree. In our ET devices resistance of 2–100 kohm have been used. Other temperature transducers employed in enzyme calorimetric analysers include Peltier elements, Darlington transistors, and thermopiles. Of these the thermistor is the most sensitive of the common temperature transducers.

Thermometry coupled to opto-acoustic analysis
The output signal from the piezoelectric transducer of the photo-acoustic spectrometer is a function of optical and thermal properties of the sample [28].

$$S = kE\alpha\beta v/C_p \tag{8}$$

where, S = transducer signal, k = constant, E_p = pulse energy, α= optical absorptivity, β =thermal expansion coefficient, v = speed of sound and C_p = specific heat.

The expression (Equation 8) shows that the signal depends on such properties of the medium as the heat capacity and the thermal expansion coefficient, which are much lower for organic solvents than for water. Relative sensitivities for photo-acoustic spectrometry performed in various organic solvents in comparison with water have been calculated. Only a very thin layer of the sample within the thermal diffusion length (μ) contributes to the signal; μ is given by

$$\mu = (\kappa/\rho C_p \pi f)^{1/2} \tag{9}$$

where κ = heat conductivity, ρ= solvent density, and f = frequency. The thermal diffusion length is somewhat shorter in organic solvents, though, because of their lower heat capacities; however, this effect is nearly cancelled by the higher heat conductivity of water.

2   Materials

Chemicals
All reagents used were of analytical grade.

Enzymes
- Catalase, EC 1.11.1.6 (Boehringer GmbH Biochimica, Mannheim, F.R.G.) Lipase preparation 7023 C, EC 3.1.1.3 (Röhm Pharma), retinol binding protein (RBP), anti-RBP, urease (type IV) (EC 3.5.1.5) and all other enzymes were from Sigma Chemical Co., St. Louis, MO, USA.
- Creatinine iminohydrolase (EC 3.5.4.21) was from Alto bioreagents, Dublin, Ireland.

Equipment
- Support: The immobilization of the enzymes glucose oxidase, catalase, peroxidase, lipase, α-chymotrypsin could be carried out on CPG, Eupergit C or Gore-Tex membrane (0.45 μ porous teflon membrane) was from W.L. Gore & Associates, Elkton, Maryland), USA.
- Conventional device: Different types of plexiglas constructions containing the immobilized enzyme column were employed initially. The earlier versions of these devices were thermostated by a water bath and later replaced by a metal block. The temperature at the exit point from the column was monitored with a thermistor. The latter was linked to a Wheatstone bridge designed for general temperature measurements and osmometry. The enzyme thermistor concept was patented in several major countries.

Such simple plexiglas devices were extremely useful and could be employed for determinations down to 0.01 mM. An example of such a device has been described in detail in several publications [29, 30]. The temperature was measured at the top of the column with a thermistor (10 kohm at 25 $^{\circ}$C, 1.5 × 6 mm, or equivalent) attached with an epoxy resin at the tip of a 2 mm (OD) acid proof steel tube. The temperature was measured as an unbalanced signal of a sensitive Wheatstone bridge. At the most sensitive setting the recorder output produced 100 mV at a temperature change of 0.01 $^{\circ}$C.

These devices were usually coupled to flow injection systems operable at flow rate of more than 1 mL min^{-1} with a low-noise peristaltic pump. The sample (0.1–1 mL) was introduced with a three-way valve or a chromatographic sample loop valve. The height of the resulting temperature peak was used as a measure of the substrate concentration and was found to be linear with substrate concentration over a wide range. Typically it was 0.01–100 mM, if not limited by the amount of catalyst/enzyme or deficiency in any of the reactants. For example, this type of instrument was adequate for the determination of urea in clinical samples [31]. The sensitivity was high enough to permit a 10-fold dilution of the samples, which eliminates problems with non-specific heat. The detection limit was consequently about 0.1 mM, and up to 30 samples could be measured per hr.

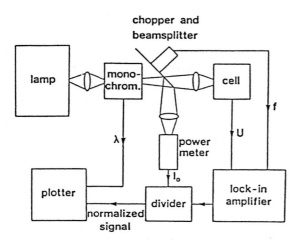

**chopper and
beamsplitter**

Figure 1 Block diagram of a
optoacoustic spectrometer
(detailed description, see text).

- Opto-acoustic device: The principle design of an opto-acoustic spectrometer [7] is shown in Figure 1. In the open cell configuration employed in our experiments, a horizontal sapphire window is connected to a pressure transducer (a piezoelectric crystal). The sapphire window is illuminated from below by chopped light from a halogen lamp. A filter with a maximum wavelength at 513 nm is placed between the lamp and the sapphire window; the chopper frequency is either 2 or 33 Hz (at low frequency there is increased penetration into the membrane of a specific thickness). A sample placed on the sapphire window absorbs light and the subsequent heat generation in the sample is transferred to the window, which expands and exerts a mechanical stress on the piezoelectric crystal. The stress is converted to a potential which is then processed to an amplitude displayed on the panel meter and registered by a recorder.

3   Methods

Protocol 1   Thermometric sensing

1. Pre-silanised (with γ-aminopropyl-triethoxysilane) CPG (controlled pore glass, pore diameter 40–80 mesh) from Pierce Chemical Co., Rockford, IL, USA were used.

2. Activate 1 g of the CPG with 2.5% glutaraldehyde in 0.01 M phosphate buffer pH 7 at room temperature (27 °C) for 2 hrs.

3. Gently shake the enzyme solution (in 0.1 M sodium phosphate, pH 7.0) with the activated CPG beads overnight at 4 °C.

4. In a given application (described later) 1 g of wet, activated CPG was treated with 1,000 IU of glucose oxidase (EC 1.1.3.4; type VII from *Aspergillus niger*) and 130,000 IU of catalase (EC 1.11.1.6; from beef liver) in 2 ml of 0.1 M sodium phosphate pH 7.0. The immobilization was carried out as in step 3.

5. In another application with β-lactamase (EC 3.5.2.6; type I from *Bacillus cereus*), 11,200 IU was immobilized per gram of activated CPG.

6. A peroxidase (EC 1.11.1.7; type VI from horseradish) column was prepared by immobilizing 2,500 IU of the enzyme to 1 g of activated CPG.

7. α-chymotrypsin (EC 3.4.21.1; type II from bovine pancreas) was immobilized by adding 1,900 IU of enzyme to 1 g of activated CPG in 2 mL of 0.1 M sodium phosphate buffer, pH 7.8.

8. Immobilization of α-chymotrypsin was also performed on Eupergit C by adding 1,900 IU of enzyme to 4 mL of 1 M potassium phosphate, pH 7.5, containing 0.05% ethyl 4-hydroxybenzoate (antibacterial agent) to 1 g of Eupergit C.

9. The mixture in [8] is kept at room temperature for 95 hrs without stirring.

10. The lipase (EC 3.1.1.3; porcine pancreas) column was prepared by adsorption of 7,500 IU of lipase dissolved in 2 mL 20 mM Tris-HCl, pH 8.0, containing 5 mM $CaCl_2$, and added to 1 g of dry Celite (Hypoflo, Supercel from BDH Chemicals, Poole, U.K.).

11. The enzyme solution in [10] was allowed to dry on the Celite under reduced pressure for 2–3 hrs.

12. The lipase-Celite preparation was wet-packed in cyclohexane saturated with buffer.

13. For further details on column packing, running the FIA etc., see [7].

Protocol 2 Opto-acoustic sensing

1. Apply 10 IU urease (EC 3.5.1.5) to 0.22 μ membrane (GSWP. Millipore, Bedford, Massachusetts), USA.

2. Cross link with glutaraldehyde (25%), 1:10 (v/v) in 0.1 M phosphate buffer pH 7 at 4 °C, for 12 hrs and quenching of unreacted sites with bovine serum albumin.

3. Sandwich membrane in (1) between 0.2 μ porous teflon membrane (FGLP, Millipore) and a spectrapor dialysis membrane (MW cut-off of 12,000–14,000, Spectrum Medical) soaked in a solution of a pH indicator, e.g. bromophenol blue in 10 mM sodium borate buffer, pH 9.0.

4. In case of creatinine iminohydrolase (EC 3.5.4.21, Aalto bioreagents, Dublin, Ireland) immobilize similar to 1–3.

5. In case of glucose oxidase 30 IU of glucose oxidase is applied to one side and 57 IU horseradish peroxidase applied to other side of Gore-Tex membrane.

6. A drop of 0.1 M sodium phosphate buffer pH 7 or an organic solvent containing phenol and aminoantipyrine is placed on the sapphire window.

7. The membrane is mounted with the glucose oxidase side up and held in place with a cylindrical weight, with an axial hole for sample addition.

8. Place the membrane package on the sapphire window of the instrument and read the amplitude after stabilisation of the signal for 30 sec.

9. Apply 5–25 μL drop of the sample on top of the membrane and record the signal between 1–3 min.

4 Troubleshooting

Unfortunately, the operational stability of the Celite columns was lower than that of the CPG columns, but the stability of the columns could in general be increased and the activity in many cases fully restored by re-conditioning with buffer. Celite was tested as a support material since it was expected that its high water-regaining capacity would stabilize the enzyme. In this study we also obtained a considerably higher activity in non-aqueous systems for lipase adsorbed on Celite than for lipase immobilized on CPG.

The development of themistor-based analysis has been slow due to its limited acceptability. The major reason being the non-discriminative detection method. However its distinct advantages include excellent operational stability (for several years), no fouling of the transducer (thermistors), ability to work in the flow injection analysis mode, negligible effects of local changes in pH and adaptability to interfacing with optical or electrochemical transducers. These have enabled simultaneous detection of analytes using multi enzyme systems [36]. In total, over 30 instruments have been assembled for applications in academics and industry.

The experience so far indicates that it is possible to work with immobilized enzymes in organic solvents over extended periods of time. Both calorimetry and opto-acoustic measurement in such media offer interesting possibilities. Usually, an increase in sensitivity can be observed, but our present knowledge is too limited to allow us to predict the effects. In some cases, additions of small amounts of organic solvents to the medium could give an unexpectedly large increase in sensitivity, mostly due to an increased enthalpy change. In other cases, the increase is much smaller and is mainly caused by the lower heat capacity of the organic solvent. In our experience, for several enzymes the activity lost during operation in organic solvents could be restored, at least partially, by re-conditioning with buffer. Furthermore, we found that enzyme columns stored in buffer between experiments could be used for several weeks for analyses in organic solvents.

5   Remarks and Conclusions

5.1   General remarks

Substantial changes in sensitivity have been observed with some of the enzymes when using pure organic solvents. Thus, the temperature response of a peroxidase column for equimolar amounts of hydrogen peroxide and hydroquinone was 45 times higher in toluene (saturated with buffer) than in buffer alone. The addition of 5% (vol/vol) diethyl ether to the toluene resulted in an even higher temperature response (50 times larger than in buffer), but further increase in ether concentrations reduced the response. Measurements of the peroxidase activity photometrically at 510 nm in toluene and in toluene with 5 and 10% (vol/vol) diethyl ether revealed that the activity was 10.6, 13.9 and 8.8 times higher, respectively, than in 0.1 M Tris-HCl buffer, pH 7.0. The increased activity resulted in sharper and higher temperature peaks, but the major contribution to the large increase in sensitivity is most likely to be a combination of an increased enthalpy change and decreased heat capacity of the solvent. Similarly the pancreatic lipase-Celite column was run with tributyrin (glyceryl tributyrate) as substrate. With 20 mM Tris-HCl, pH 8.0 containing 5 mM $CaCl_2$ and 4% Triton X-100; An 8.5 m^{o}C temperature peak was obtained for 50 mM tributyrin. The peaks were rather broad but without any trailing. The peaks in cyclohexane were broad with a slightly trailing end.

In the case of opto-acoustic measurements, there was a considerable gain in sensitivity from measuring the chromophores in an organic solvent medium. By comparing the response of some dyes in water and organic solvents, the sensitivities in the latter were found to be about 10–15 times higher e.g. bromothymol blue in toluene and carbon tetrachloride. Larger gains in sensitivity could be expected by operating in supercritical fluids such as, supercritical CO_2. However such measurements may require redesigning the present set-up for opto-acoustic measurements.

Increased solubility of reactants, including gases such as oxygen, in organic solvents is an additional advantage. The possibility for new reaction routes and even reversed reaction direction of enzymes working in organic solvents was exemplified by the studies on peptide synthesis using α-chymotrypsin. Despite some obvious drawbacks associated with the work in organic solvents, such as need to find suitable solvent-resistant construction materials there would be many situations in which it is advantageous to operate thermal/opto-acoustic biosensors (enzyme thermistors, ETs) in organic solvents.

5.2   Glucose biosensor

In a study of the organic phase effects on bio-thermal analysis using enzyme thermistors, we reported [30] the influence of varying amount of ethanol (1–10 mol % or 3–27 vol/vol) on the temperature response of a glucose oxidase/catalase column using 0.5 mL glucose samples (in the FIA mode) at a flow rate of 1.1 mL/min. The temperature response was found to increase with ethanol added to the buffer up to 2 mol % at which point it was 2.33 times higher than with pure phosphate buffer. At 10 mol % ethanol, the sensitivity was 1.84 times higher and at higher concentrations (>15 mol % or 37% vol/vol) the sensitivity became lower than with pure buffer. This decrease in sensitivity could be attributed to the inactivation of the enzymes by the ethanol. In order to study whether the enzymes responded differently to the ethanol addition, 0.5 mM glucose or 1 mM hydrogen peroxide samples were run on a glucose oxidase/catalase column in buffer/ethanol mixtures from 0 to 99.5 mol % ethanol. The results indicated that the immobilized glucose oxidase activity was considerably lower than that of catalase. The temperature response of the catalase reaction for 1 mM hydrogen peroxide was about twice as large as that for 0.5 mM glucose in the combined reactions at lower ethanol concentrations. Based on the reaction enthalpies one could predict a similar response. However, catalase seem to be more prone to denaturation with increasing concentration of the organic solvent. At higher ethanol concentrations, the response of GOX and Catalase were about the same, and the maximum response occurred at the same ethanol concentration for both glucose and hydrogen peroxide. In fact when other alcohols were tried, the order of temperature response was observed to be 1-butanol < 2-butanol < buffer < 1-propanol < methanol < ethanol < 2-propanol.

Glucose determinations with the opto-acoustic technique at 2 Hz resulted in a signal of 79 mV for 100 mM glucose and 24 mV for 10 mM glucose with carbon tetrachloride. The corresponding figures for buffer were 44 mV and 15 mV, respectively (Fig. 2). Similarly the sensitivity of urease membrane was 235 mV for 600 mg/mL (10 mM) urea and 142 mV for 100 mg/mL at 2 Hz with the indicator dissolved in the buffer.

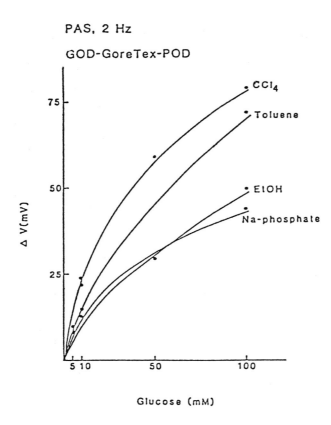

PAS, 2 Hz

GOD-GoreTex-POD

Figure 2 Calibration curves for glucose obtained with glucose oxidase and peroxidase bound to a Gore-Tex membrane in an optoacoustic spectrophotometer with a 2-Hz chopper frequency.

The phenol and aminoantipyrine used for the peroxidase reaction were dissolved in various solvents (carbon tetrachloride, toluene, ethanol) and in 0.1 M sodium phosphate buffer, pH 7.0. The experiments were carried out at 25 °C.

5.3 Peroxide biosensor

Recently, both hydrogen peroxide and butanone peroxide were estimated in aqueous and organic milieu respectively (unpublished data). With increasing interest in the use of pure organic solvents, attempts were made to employ a suitable organic phase with water content between 1–1.5% (vol/vol). While in previous investigations, hydrogen peroxide had been employed with organic solvents, in the present instance butanone peroxide, a more favoured substrate for HRP in organic phase was employed [32]. At the outset it is clear (Fig. 3) that the sensitivity of butanone peroxide detection in acetone (with 1% aqueous phase) was higher than hydrogen peroxide in similar concentrations in aqueous phase. Employing a variety of other organic phase such as methanol,

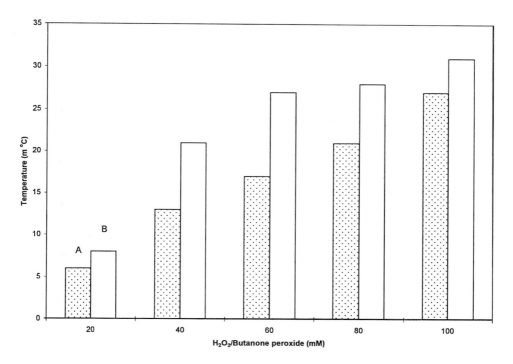

Figure 3   A comparison of the thermometric response to varying concentrations (20–100 mM) of H_2O_2 and butanone peroxide using HRP immobilized on CPG and packed in a 1 mL column, inserted into a thermal assay probe.

The mobile phase (in a flow injection system) is acetone (1% water) at flow speed of 1 mL min[-1]. A ≡ H_2O_2 in aqueous phase B ≡ Butanone Peroxide in organic phase

acetonitrile and butanol, it was observed that the increase in response was in the order of 2-butanol< acetonitrile < methanol < acetone (Fig. 4). The reproducibility for a CPG-immobilized HRP column was excellent with ±7% (CV %). The presence of o-phenylene diamine (OPD) as a hydrogen donor for the HRP reaction had a negligible effect on the overall performance of the column, as the HRP activity immobilized on such columns was much higher and the presence of OPD infact had a negative influence on the sensing response in acetone (Fig. 5). The influence of varying amounts of the aqueous phase was also tested with methanol as the solvent. Figure 6 illustrates the effect of 10%, 20% and neat methanol (1% water) on the response of the thermistor. The last was found to provide the most favourable response compared with the presence of 10 and 20% water. In all these instances the linearity was found to be between 10–70 mM butanone peroxide which provides a good working range, although in most cases non-linearity set in thereafter. A notable feature in the present investigations was the nature of the response, i.e. the thermometric response peaks were much sharper and well defined in the nonaqueous phase compared to the aqueous phase, and a plot of the FWHM (Fig. 7) indicates that in

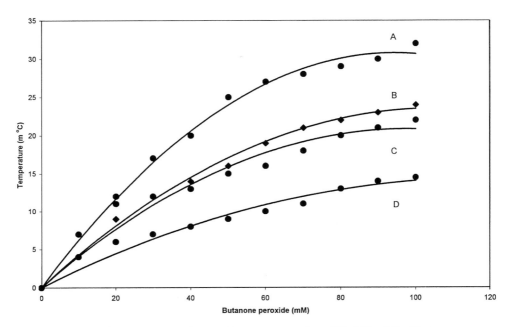

Figure 4   The calibration curves for varying concentrations (10–100 mM) of butanone peroxide using acetone (A), methanol (B), acetonitrile (C) and 2-butanol (D), using HRP immobilized on CPG and packed in a 1 mL column, inserted into a thermal assay probe.

A-D contain 1% water. Flow speed of 1 mL min^{-1}.

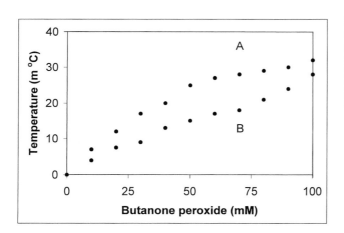

Figure 5   The effect of 0.5 mM o-phenylene diamine (curve B) on the response of HRP to varying concentrations (10–100 mM) of butanone peroxide, compared to its absence (curve A).

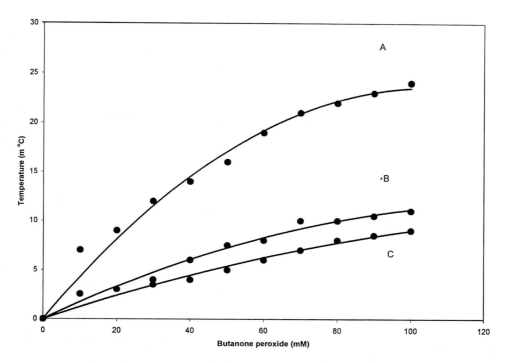

Figure 6   The effect of varying water % (A) neat methanol (1% water) (B) 10% water and (C) 20% water on the response of HRP to varying concentrations (10–100 mM) of butanone peroxide.

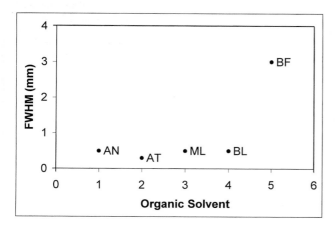

Figure 7   The width at half maximum (FWHM) of the thermometric peak obtained for sensing of butanone peroxide by HRP using the following organic solvents: acetonitrile (AN), acetone (AT), methanol (ML) and 2-butanol (BL) compared to FWHM in phosphate buffer (BF) 0.1 M pH 7 for H_2O_2.

the organic phase the FWHM is almost three times lower than in aqueous phase. This indicates the effect of increased enthalpy change coupled to the lower heat content of organic phase compared to the aqueous phase. Based on these investigations it could be suggested that the thermometric technique is suitable for rapid screening of enzyme activity in various organic solvents.

As an extension of this approach we explored the possibility of studying the retinol and retinol binding protein (RBP) complex using HRP as a tag on retinol. The experiments were aimed at understanding the stability of the retinol-RBP complex using an HRP-retinol conjugate. Table 2 summarises the results with different solvents used to challenge the RBP-retinol-HRP complex. The solvents chosen had varying hydrophobicity and log P values (unpublished data).

Table 2 A list of various organic solvents and combinations attempted to study the RBP-retinol complex using HRP as a marker enzyme.

RBP (retinol binding protein) was immobilized on controlled pore glass beads using glutaraldehyde. The beads were packed in a column. Retinol-HRP conjugate was injected into the column resulting in a RBP-retinol – HRP complex. Different solvents /solutions were injected to test the stability of RBP-retinol bond. The dissociation is reflected in the decrease in the signal upon injecting hydrogen peroxide (because of the presence of HRP tag).

Screening number	Solvent system	Concentration	Thermometric peak-height (mm)
1	Phosphate buffer	0.1 M, pH 7.0	23
2	β-mercaptoethanol	Neat	15
3	Ethanol	Neat	15
4	Ammonium sulfate + ethanol	0.1 M + 50%	17
5	Ethylene glycol	Neat	16
6	Sodium chloride	0.1 M	17
7	Dimethylformamide	Neat	17
8*	Phosphate buffer	0.1 M, pH 7.0	6
9*	Glycine buffer	0.1 M, pH 2.2	6
10	Chloroform + Methanol	Neat	26
11	Phosphate buffer	0.1 M, pH 7.0	23

* The output corresponds to 20 mM injection. For all other solvents 80 mM hydrogen peroxide was used.

5.4 Penicillin biosensor

Penicillin was determined with a β-lactamase (penicillinase) column. The addition of 5% (vol/vol) ethanol as co-solvent in 0.1 M Tris-HCl buffer, pH 7.0, resulted in 2.3 times higher response for 10 mM penicillin G than in buffer alone, at higher ethanol concentrations (50%); on the other hand, the increase in sensitivity was much lower (Fig. 8).

$$\text{Penicillin} + H_2O \rightarrow \text{Penicilloate} \tag{10}$$

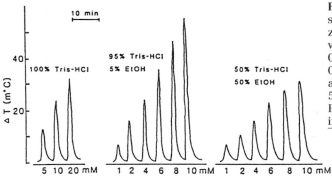

Figure 8 Temperature responses obtained by an enzyme thermistor equipped with a β-lactamase column for 0.5 mL penicillin G samples in 0.1 Tris-HCl buffer, pH 7.0, and in the same buffer with 5% and 50% ethanol added. Flow rate: 1.0 mL min^{-1}, working temperature 30 °C.

5.5   Biosensors in peptide synthesis

Immobilized α-chymotrypsin had been employed to monitor the hydrolytic and synthetic reactions [33, 34]. A direct correlation was demonstrated between the negative ΔT values registered and the amount of peptides formed (essentially various N-acetyldipeptide amides), allowing concentrations of 0.1 mM peptide and lower to be determined directly in the reaction medium. The thermometric response for the hydrolytic reactions with various ester substrates was tested with different buffer and solvent systems to find optimal conditions. It was shown that esterolytic reactions can be conveniently assayed using the ET and that low concentrations of organic solvents do not interfere significantly. We chose to perform determination of esters in 50 mM Tris-HCl pH 7.8 containing 10% DMF (dimethylformamide). As seen from the upper part of Figure 9, a linear temperature response was found on increasing concentrations of amino acid esters. We also found that the total enzyme activity of the Eupergit columns was higher than that for CPG at the higher pH values used in synthesis. Consequently Eupergit C was used as matrix in subsequent studies.

Synthetic reactions were also followed using a similar approach. Peptide synthesis was used as a model. A solvent system consisting of 50% DMF+50% 0.1 M sodium borate, pH 10 was selected. The lower portion of Figure 9 shows a linear correlation of the signal with increasing amounts of the amino acid esters (the amide, Leu-NH$_2$, was kept constant), this time due to the endothermic nature of peptide formation, resulting in 'negative ΔT values'. Nearly the same temperature responses were obtained in the formation of peptides using other amides.

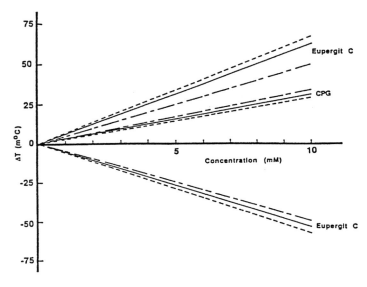

Figure 9 Temperature responses obtained with an ET containing immobilized α-chymotrypsin.

The lower part of the figure (negative temperature peaks represents results of dipeptide synthesis from Ac-Tyr-OEt (— – —), Ac-Phe-OEt (– – – –) and Ac-Phe-OMe (——) combined with Leu-NH$_2$ in 50% DMF + 50% 0.1 M sodium borate, pH 10.0. In the upper part of the figure the responses obtained on the hydrolysis of the same esters are shown for α-chymotrypsin immobilised on CPG and Eupergit C, respectively, using 10% DMF in 0.05 M Tris-HCl, pH 7.8.

5.6   Fatty acid biosensors

A pancreatic lipase-Celite column was run with tributyrin (glyceryl tributyrate) as substrate [5]. With 20 mM tris-HCl, pH 8.0 containing 5 mM CaCl$_2$ and 4% Triton X-100, an 8.5 m°C thermometric peak was obtained for 50 mM tributyrin. Higher sensitivity (2.1 times) was obtained (10 m°C for 50 mM tri-butyrin) with cyclohexane saturated Tris-buffer as solvent. This effect could be expected from the respective heat capacities for the two media. The excellent solubility of the lipase substrates in cyclohexane [35] is a major advantage that considerably facilitates the assay. An investigation on cholesterol has also been recently reported [36].

Abbreviations

ET	Enzyme thermistor
CPG	Controlled pore glass
OPB	Organic phase biosensor
TEP	Thermal enzyme probe
ID	Internal diameter
OD	Outer diameter
FIA	Flow injection analysis
MW	Molecular weight

IU   Internal unit (enzyme activity)
GOD   Glucose oxidase
HRP   Horse-radish peroxidase
FWHM   Full width at half maximum
DMF   Dimethyl formamide
ΔT Change in temperature
RBP   Retinol binding protein

References

1   Clark LC, Lyons C (1962) Electrode systems for continuous monitoring in cardiovascular surgery. *Ann NY Acad Sci* 102: 29–45

2   Turner APF, Karube I, Wilson GS (1987) Biosensors: Fundamental & Applications. Oxford University Press, Oxford

3   Wise DL, Wingard LB (1991) Biosensors with Fibreoptics. Humana Press, NY

4   Gupta MN (1992) Enzyme function in organic solvents. *Eur J Biochem* 203: 25–32

5   Flygare L, Danielsson B (1988) Advantages of organic solvents in thermometric and optoacoustic enzyme analysis. Ann NY Acad Sci 542: 485–496

6   Hall GF, Turner APF (1991) An organic phase enzyme electrode for cholesterol. *Anal Lett* 24: 1375–1388

7   Danielsson B, Flygare L (1990) Performance of a thermal-biosensor in organic solvents.*Sens Act B* 1: 523–527

8   Saini S, Turner APF (1991) Biosensors in organic phases. *Biochem Soc Trans* 19:28–31

9   Schubert F, Saini S, Turner APF, Scheller F (1992) Organic phase enzyme electrodes for the determination of hydrogen peroxide and phenol. *Sens Act B 7: 408–411*

10 Hall GF, Best DJ, Turner APF (1988) Amperometric enzyme electrode for the determination of phenols in chloroform. *Enzyme Microb Technol* 10: 543–546

11 Wang J, Lin Y, Chen L (1993) Organic-phase biosensors for monitoring phenol and hydrogen peroxide in pharmaceutical antibacterial products. *Analyst* 118: 277–280

12 Weetall HH (1991) Antibodies in water immiscible solvents immobilized antibodies in hexane. *J Immunol Meth* 136:139–142

13 Mionetto N, Marty JL, Karube I (1994) Acetylcholinesterase in organic solvents for the detection of pesticides: Biosensor application. *Biosens Bioelectron* 9:463–470

14 Spink C, Wadso I (1976) Calorimetry as an analytical tool in biochemistry and biology. *Methods Biochem Anal* 23: 1–159

15 Mosbach K, Danielsson B (1974) An enzyme thermistor. *Biochem Biophys Acta* 364: 140–145

16 Danielsson B, Mosbach K (1988) Enzyme thermistors. In: K Mosbach (ed): *Methods in Enzymology*. Academic Press Inc., London, Vol no. 137, 181–197

17 Ramanathan K, Khayyami M, Danielsson B (1998) Enzyme biosensors based on thermal transducer/ thermistor. In: A Mulchandani, KR Rogers (eds): *Methods in Biotechnology: Enzyme and Microbial biosensors: Techniques and protocols.* Humana Press Inc., Totowa, NJ, Ch 13,175–186

18 Ramanathan K, Khayyami M, Danielsson B (1998) Immunobiosensors based on thermistors. In: A Mulchandani, KR Rogers (eds): *Methods in Biotechnology: Affiniy Biosensors : Techniques and Protocols.* Humana press Inc., Totowa, NJ, Ch 2, 19–29

19 Xie B, Ramanathan K, Danielsson B (1999) Principles of enzyme thermistor systems: Applications to biomedical and other measurements. In: T Scheper (eds): *Advances in biochemical engineering /biotechnology*. Springer Verlag, FRG Vol 64, 1–33

20 Xie B, Mecklenburg M, Danielsson B et al. (1994) Microbiosensor based on an integrated thermopile. *Anal Chim Acta* 299: 165–170

21 Xie B, Danielsson B (1996) An integrated thermal biosensor array for multianalyte determination demonstrated with glucose, urea and penicillin. *Anal Lett* 29: 1921–1932

22 Wollenberger U, Schubert F, Scheller F et al. (1987) A biosensor for ADP with internal substrate amplification. *Anal Lett* 20: 657–668

23 Salman S, Haupt K, Ramanathan K, Danielsson B (1997) Thermometric sensing of fluoride by adsorption on ceramic hydroxyapatite using flow injection analysis. *Anal Commun* 34: 329–332

24 Iwuoha EI, Smyth MR (1993) Effect of organic solvents on the behaviour of a glucose oxidase-based biosensor. In: G Guilbault, M Mascini (eds): *Uses of immobilized biological compounds.* Kluwer Academic publication, Doretcht, 245–254

25 Iwuoha EI, Smyth MR(1994) Reactivity of a glucose oxidase electrode in polar organic solvents *Anal Proc* 31: 19–21

26 Iwuoha EI, Smyth MR (1994) Organic phase application of amperometric glucose sensor. *Analyst* 119: 265–268

27 Product catalogue Victory Engineering Corporation, Victory Road, Springfield, New Jersey 07081, USA

28 Hardcastle F, Leach R, Harris J (1986) Photoacoustic spectroscopy in supercritical fluids. *Anal Chem* 58: 493–496

29 Danielson B, Mattiasson B (1996) Thermistor- based biosensors. In: RF Taylor, JS Schultz (eds): *Handbook of chemical and biological sensors.* IOP publishing Ltd., Ch 20, 495- 511

30 Danielsson B, Flygare L, Velev T (1989) Biothermal analysis performed in organic solvents. *Anal Lett* 22, 1417–1428

31 Xie B, Harborn U, Mecklenburg M, Danielsson B (1994) Urea and lactate determined in 1μL whole-blood samples with a miniaturized thermal biosensor. *Clin Chem* 40: 2282–2287

32 Iwuoha EI, Smyth MR, Lyons MEG (1997) Organic phase enzyme electrodes: Kinetics and analytical applications. *Biosens Bioelectron* 12: 53–75

33 Stasinska B, Danielsson B, Mosbach K (1989) The use of biosensors in bioorganic synthesis: Peptide synthesis by immobilized α-chymotrypsin assessed with an enzyme thermistor. *Biotechnol Tech* 3: 281–288

34 Adlercreutz P, Clapes P, Mattiasson B (1988) Enzymatic peptide synthesisin organic media. *Ann NY Acad Sci* 613: 517–520

35 Zaks A, Klibanov AM (1985) Enzyme-catalyzed processes in organic solvents. *Proc Natl Acad Sci USA* 82: 3192–3196

36 Raghavan V, Ramanathan K, Sundaram PV, Danielsson B (1999) An enzyme thermistor-based assay for total and free cholesterol. *Clin Chim Acta* 289: 145–158

12 Importance of the Medium for *in vitro* and *in vivo* Protein Folding Mechanisms: Biomedical Implications

Ajit Sadana

Contents

1 Introduction

Protein folding is a central problem in biology, and has intrigued biologists for quite some time. During protein folding the protein acquires its native and active state starting from a linear inactive state, where the information for its coding is contained in its amino-acid sequence. Besides the bioprocessing implications of correct folding, this problem has acquired increasing importance and attention recently due to its implication in some persistent diseases that are presently incurable or against which the progress and the subsequent cure is painfully slow. Different authors [1–3] have emphasized that for the correct protein folding *in vivo*, molecular chaperones are required. This is the case also for *in vitro* studies. Hartl [4] emphasizes that if this pre-existing machinery is not functioning properly, if mutations exist, or for other reasons, then proteins will misfold. The misfolding of proteins will subsequently lead to the build-up of aggregates either in living tissue (*in vivo)* or in bioprocessing solutions (*in vitro*). The build-up of aggregates in living tissue has been identified as a cause for many intractable diseases such as Alzheimer's. It is these mis-

Methods and Tools in Biosciences and Medicine
Methods in non-aqueous enzymology, ed. by M. N. Gupta
© 2000 Birkhäuser Verlag Basel/Switzerland

folded proteins that give rise to senile plaques and neurofibrillary tangles [5] that lead to the commonest form of dementia, Alzheimer's disease. Jen et al. [6] indicate that the peptide, amyloid-β peptide (Aβ) seems to play a leading role in the onset of the disease. They also add that the toxicity of this peptide to neurons has already been shown *in vitro*. More recently, the mutations in the tau protein have been confirmed as neuron killers [7]. It is apparently the major component of the aggregated proteins found in brain cells. Apparently, the severity of the dementia may be correlated with the number of tangles or degree of aggregation (misfolding). Borman [8] indicates that though chaperones play a key role in the refolding process, there is still a great deal of uncertainity involved. But, chaperones do provide an avenue for therapeutic intervention, for example, against prion infection. Hence, the recent and ever-increasing importance for promoting an understanding of how proteins fold, and how they may be made to fold correctly. Protein folding occurs in both aqueous and non-aqueous environments. Here we intend to emphasize *in vivo* protein folding/unfolding which occurs generally under nonaqueous conditions considering the complexities involved and the cellular environment where the protein folding takes place.

Mitraki and King [9] emphasize that due to the flexibility of the polypeptide chains a large number of conformational states are possible. The selection of the correct route of the native active state, and the avoidance of the several incorrect routes that lead to the misfolded forms is the challenge that confronts present day protein chemists. A new view of protein folding has emerged that postulates that there are many ways to reach the native and active structure. Lazardis and Karplus [10] indicate that in the new view of protein folding there is a general bias of the energy surface towards the native state. This would help in the search for the correct folding pathway since a specific pathway is not required. Chan [11] indicates that theoreticians analyze the folding of generic proteins, whereas experimentalists deal with specific proteins. It is often difficult to reconcile their separate views. Pain [12] has succinctly indicated that protein folding is one of the most intriguing intellectual challenges in molecular biology.

Information available in the open literature regarding *in vivo* protein folding will be presented with particular emphasis on aggregate formation, and how it may be minimized. *In lieu* of any information presented to the contrary, it is assumed that *in vivo* protein folding/unfolding exhibits similar characteristics under aqueous and nonaqueous environments. If one were forced to make a distinction then one can presume that a large number of *in vitro* protein unfolding/folding reactions occur in aqueous solution, and one can with caution extend these basic characteristics to *in vivo* mechanisms occurring under nonaqueous conditions. We will be drawing upon the information available for *in vitro* folding mechanisms to help understand *in vivo* folding. This is because information on *in vivo* protein mechanisms is relatively scarce.

During the formation of proteins using DNA technology, inclusion bodies are formed. These inclusion bodies are intracellular aggregates or refractile bodies that consist of densely packed protein molecules that have partial secondary structure [13]. This aggregative behavior of proteins is more serious when the deposition occurs in living tissue. These amyloid deposits in brain tissue often lead to different diseases such as the common memory disorder, Alzhemier's [14]. Borman [8] indicates that other fatal neurodegenerative disorders characterized by deposits of protein in the brain include Creutzfeld-Jakob disease (CJD), and bovine spongiform encephalopathy (BSE), or mad cow disease.

Proteins have a hierarchical structure; during protein folding subdomains are initially formed. These subdomains then combine with other domains to produce the final active structure of the protein. This process involves a lot of similar (though not identical) repeating biochemical units. Even in the complex pattern structure there is a repeating pattern. It may be suggested that it is this repeating pattern and the characteristic heterogeneity observed in protein structure that may be described by fractals. Fractals are disordered systems, and the disorder is described by nonintegral dimensions [15]. These authors indicate that as long as surface irregularities show scale invariance that is dilatational symmetry they can be characterized by a single number, the fractal dimension. Surfaces of progressively higher irregularity are characterized by progressively higher order fractal dimensions. A characteristic feature of fractals is the self-similarity observed at different levels of scale. It would seem appropriate to represent the different folding stages using a fractal analysis. This is because of the inherent repeating pattern or similar structures formed during the folding stages.

We will attempt to analyze processes by which correct *in vivo* folding is enhanced and incorrect or misfolded protein folding is minimized. For example, Hartl [4] emphasizes that chaperones bind to an unstable protein conformer and facilitate its attainment of the correct form. They do increase the yield, but not the rate of folding reactions. Folding rate increases are caused by folding catalysts such as protein disulfide isomerases and peptidyl-proyl isomerases. Schmid [16] indicates that these enzymes actually increase the rate of some slow steps in some proteins. Qualitative fractal dimension ranges (that may be determined by diagnostic or other methods) of the misfolded and the correctly folded protein are required to help provide an independent measure of the correct and misfolded forms of the protein. An increase in the fractal dimension indicates an increase in the heterogeneity or the degree of inhomogeneity or the surface roughness. This review should assist one in developing effective strategies to help eliminate or minimize the formation of aggregates. This should then lead to a possible cure of some of these intractable diseases.

2   Misfolding of proteins

Protein folding is constrained both by kinetics as well as by thermodynamics [17]. The kinetic nature arises due to the vectorial nature of the protein synthesis, and the thermodynamic nature arises due to the necessity of energy minimization of the different states. The driving force of the three-dimensional protein structure formation is the minimization of the free energy of stabilization. There is a hierarchical nature of the three-dimensional structure formation [18]. Short-ranged interactions lead to secondary structures. These secondary elements, through the process of gradual combination and reshuffling, lead to the formation of subdomains, domains, and subunit assemblies. Note the modular nature of the three-dimensional structure formation. This modular nature is very amenable to fractal analysis, which is based on self-similarity of some basic units.

In vitro folding is often constrained by competing processes [9, 19]. Aggregation may occur due to the association of hydrophobic surfaces that are exposed in folding intermediates [20]. Efforts are being made to minimize these aggregative interactions. Because *in vivo* folding of proteins occurs successfully in a complex milieu of reactions, there are indications that a special class of *in vivo* proteins have the capability of directing or facilitating protein folding reactions. These proteins are called chaperonins [1]. They accomplish this *helping* or *aiding* task, in part, by interacting with partially folded intermediates of target proteins. These chaperonins are quite specific in the sense that they have a preference for certain amino-acid patterns [4]. Thus, they will interact (or bind to) a wide range of unfolded polypeptides, while completely ignoring the native or active states of the protein. The interaction of the chaperonins with the partially unfolded proteins minimizes the interactions of the hydrophobic surfaces that lead to the aggregation of proteins. *In vitro* folding aggregation can be controlled to some extent by dilution of the protein concentration. Molecular crowding, however, prevents this *in vivo* folding. This leads to higher values of the association constants. It is then left to the chaperones to help minimize these interactions or aggregative effects [4]. This author emphasizes that the chaperones play quite a few additional roles that include their involvement in protein metabolism under stress as well as non-stress conditions, membrane translocation, degradation of misfolded proteins, and regulatory processes.

Now let us try to tie-in the protein folding mechanism with fractal concepts. The folding step may be summarized by a multistep, mechanistic scheme [21]. This is consistent with the self-repeating nature inherent in fractal concepts, where the sub-structures exhibit self-similarity on different levels of scale. This is consistent with the merging of the individual subdomains and hierarchical structure components and the consecutive folding process. These concepts are also consistent with the simple and elegant cardboard-box model for the protein unfolding/folding transition(s) [22].

Kotlarski et al. [23] have proposed the following model for refolding. It involves the very basic and essential steps required. These steps are:

$$\text{unfolded} \rightarrow \text{intermediate} \rightarrow \text{native}$$
$$\downarrow$$
$$\text{aggregate}$$

This basic series-parallel mechanism is consistent with most other proposed mechanisms of protein refolding. Note that the aggregate is formed in the *off-pathway* step. Ghag et al. [24] have proposed a refolding pathway for *Cephalosporium acremonium* from granules of recombinant *E. coli*. This mechanism involves a few more steps than the basic mechanism proposed above. This mechanism is:

$$\text{Unfolded} \rightarrow \text{Intermediate} \rightarrow \text{Refolded}$$
$$\text{(Monomer)} \quad \text{(Monomer)} \quad \text{(Monomer)}$$
$$\downarrow$$
$$\text{Dimerization}$$
$$\downarrow$$
$$\text{Polymerization}$$
$$\downarrow$$
$$\text{Irrecoverability/Degradation}$$

These authors indicate that the main obstacle for obtaining a correctly folded enzyme is a urea-dependent aggregation that leads to irreversible enzyme inactivation. The critical step is to stabilize the monomeric form under appropriate environmental conditions, which then leads to large quantities of highly pure, active, and stable enzyme.

Inclusion bodies (IB) are intracellular protein aggregates or refractile bodies. Inclusion bodies are formed from partially folded intermediates and not from completely unfolded protein [25]. There is a competition between the first-order (correct folding reaction) and the diffusion-controlled higher-order reaction which may be described by:

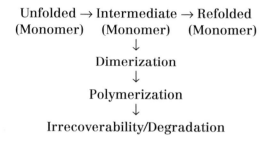

$$U \xrightarrow{k_1} N$$
$$k_2 \downarrow$$
$$A$$

Here U, N, and A are partially folded intermediates, and the native and aggregated states, respectively. k_1 and k_2 are first-order folding and second-order aggregation rate constants, respectively. Heterogeneity of the initial enzyme (in this case polypeptide chain state) plays a significant role in enzyme deactivation kinetics [26]. Presumably, heterogeneity of the initial polypeptide state will play a significant role in refolding kinetics. One may reasonably presume that because heterogeneity tends to make reaction orders higher in enzyme deactivation kinetics [27] the aggregative reaction will be preferred in refolding kinetics.

Finally, recent *in vitro* efforts to minimize aggregate formation are worth mentioning. The basic idea over here being to eliminate or minimize the hydrophobic interactions that lead to the formation of the aggregates. Some of the techniques that have been used to separate the intermediate species include (a) size-exclusion chromatography [28], hollow-fiber membrane dialysis [29], reverse micelles utilizing a solid-liquid extraction technique [30], and the utilization of protein disulfide isomerase (PDI) and prolyl isomerase (PPI) immobilized on cross-linked agarose beads [31]. Middleberg [32] has emphasized that the amount of protein recovered is very sensitive to the operating conditions. Different co-solvents have been utilized to enhance protein folding.

2.1   Effect of the medium and the co-solvent on protein folding

Cleland and Wang [33] have analyzed the effects of the co-solvent on the refolding and aggregation of bovine carbonic anhydrase (CAB). These authors indicate that these co-solvents, like sugars, enhanced the hydration of the CAB and inhibited the aggregation step, though they did not improve the recovery of the active form of the protein. The authors analyzed the influence of sugars, polyamino acids, polyethylene glycol (PEG), and PEG derivatives on the recovery and refolding of CAB, besides studying their influence on the aggregating effects. The refolding pathway contains more than one intermediate, and these authors indicate that co-solvents inhibited the aggregating effects by reversibly binding to the first hydrophobic intermediate. They proposed a thermodynamic criterion for optimum refolding of a protein in the presence of a cosolvent, and emphasized that the influence of these molecules on protein refolding should also be analyzed.

The hydrophobic effect has often been utilized to enhance protein refolding. This comprises of (a) the hydrophobic hydration which is the hydration of an apolar species in water, and (b) the hydrophobic interaction, which is the bringing together of two or more species when both are dissolved in water. Walshaw and Goodfellow [34] analyzed the distribution of solvent molecules around apolar side-chains in protein crystals. They examined the distribution of solvent sites within 5 Å of the nonpolar side chains of alanine, valine, leucine, isoleucine, and pheylalanine based on data taken from the high-resolution structures. They noted a clustering of solvent molecules into specific regions which is of a non-random nature. For example, their analysis shows that for alanine and phenylalanine the water molecules that are not within the hydrogen bonding distance of protein molecules also cluster into specific regions. They suggest that the hydrophobic hydration in protein crystals is correlated with nonpolar groups. The local environment and stereochemistry of the apolar atoms plays a significant role. Finally, this clustering of the water molecules around the solvent sites in a non-random fashion is a characteristic of the fractal nature of this system.

Powers et al. [35] have analyzed the formation of hydrophobic ion pairs between sodium dodecyl sulfate (SDS) and various peptides and proteins. They examined the solubility, stability, and structures of these biomolecules in nonaqueous solvents. These authors describe a technique for solubilizing peptides and proteins in organic solvents. The utilization of an anionic detergent such as SDS leads to diminished aqueous solubility in organic solvents. Therefore, these authors note that partitioning into nonpolar solvents may be increased by two to four orders of magnitude. For example, the solubility of the SDS-insulin complex increases by a factor of ten in 1-octanol when compared with its solubility in water. Furthermore, the native structure of insulin remains intact, and the proteins dissolved in nonpolar solvents demonstrate high stability characteristics. Powers et al. [35] emphasize that the peptides and proteins may be extracted back into the aqueous phase. This back extraction is possible by replacement of the SDS counterion by chloride.

McMinn et al. [36] have analyzed the hydration of proteins in nearly anhydrous organic solvent suspensions. These authors wanted to analyze the role water plays in dilute aqueous solution on the catalytic activity of enzymes. Their results indicate that nonpolar and moderately polar solvents act as an inert phase for water sorption by protein powders in organic suspensions. They emphasize that it would be of interest to measure the hydration level of the enzymes and relate it directly to the catalytic activity. This they suggest would permit further insights into the influence of water in these systems.

If one may carefully extend these *in vitro* results to *in vivo* studies, then one can understand the very significant role that a nonaqueous medium plays in helping to maintain the correct structure and in protein folding. It would be of interest to elucidate the medium and environment present during *in vivo* folding. Furthermore, one can understand the very significant role that chaperones play in helping to fold proteins correctly, and the significant impact that would, and does arise, when these chaperones and other agents do not function properly, and proteins misfold. In other words, it is apparently critical that these chaperones function at an optimal level. The body does have checks and balances and feedback control systems in place, and some of these may get involved and help mitigate the loss of chaperone action to some extent. Finally, in an analysis of chaperone dynamics and its influence on protein secretion, Robinson and Lauffenberger [37] indicate that the chaperone interaction with the protein structural form should be of a tight but a transient nature. They indicate that there is a decrease in the product yield, if the protein-chaperone interaction is prolonged.

3 Corrective action in cells

We now present different mechanisms by which correct folding is enhanced in cells, and the presence of misfolded forms is minimized.

3.1 ATP energy dependence

The endoplasmic reticulum (ER) provides a unique environment wherein protein folding can take place [38]. The authors indicate that the ER provides an oxidizing environment which facilitates the formation of disulfide bonds. The proteins require energy to help keep them in their oxidized and proper folded state. This energy requirement involves BiP (a chaperone). This hsp (heat shock protein) 70 analogue [39] apparently binds and hydrolyses ATP as it associates and dissociates from peptides and proteins [39, 40]. This is assumed to be a folding factor in the ER [41].

3.2 Mechanism of action of GRP 78 and GRP 94

Gething et al. [3] suggest that there is a link between the glucose-related proteins (GRP) and mutant or misfolded polypeptides. This hypothesis was tested using the misfolded forms of haemaglutinin (HA) that accumulate in the ER [42]. Gething et al. [3] indicate that these GRPs are not significantly induced by high temperature, unlike the heat shock proteins (hsp). There are indications that the GRP 78, an immunoglobulin heavy chain binding protein (BiP) binds transiently to the wild-type proteins, but in a more permanent sense to the misfolded proteins that accumulate in the ER [43]. This is consistent with optimum binding proposed by Robinson and Lauffenberger [37], as indicated previously above. Malfolding *per se* is the primary reason for the induction of the glucose-regulated proteins, GRP 78 and 94 [42]. These authors indicate that it is the malfolded forms of the HA that accumulate in the ER that are responsible for the induction of GRP 78 and 94. They emphasize that there must be a mechanism that monitors what is going on in the lumen of the ER, and must send (or transduce) signals across the ER to the nucleus. They suggest that perhaps it is the binding of the GRP to the malfolded forms, and the corresponding reduction in the free GRP that presumably leads to the stress response of the nucleus, and the eventual induction of the GRP synthesized. It is the abnormal glycosylation that may be the result of different factors that leads indirectly to the induction of the GRPs. This occurs according to the authors by the prevention or delay of the folding and the assembly of nascent polypeptides that lead to the accumulation of misfolded polypeptides in the ER.

3.3   Influence of the pro region

The protein folding reaction has been analyzed under kinetic control [44]. These authors emphasize that there is a distinct difference between the action of the pro region and that of the chaperones, as indicated above. The chaperones reduce the rate of aggregate formation (off-path way step), whereas the pro region accelerates the rate-limiting step in the folding pathway by a factor of about 10^7. They analyzed the folding of the α-lytic protease. A 166 amino-acid region is required for the correct folding of the protease domain [44–46]. Baker et al. [44] omitted the pro region, and were able to trap an inactive, but folding competent state (I) with native-like structure. Furthermore, when the pro region is added, the I state folds rapidly to the active and native form. In the absence of the pro region Baker et al. [44] estimate that the free energy barrier for conversion of the I state to the native state is in excess of 27 kcal/mol. The pro region according to these authors lowers the free energy barrier and provides access to the new regions of conformational space, that is the enzymatically active state. In the presence of the pro region, the kinetics of folding is simple and follows:

$$I + Pro \leftrightarrow I.Pro \rightarrow Native\text{-}Pro$$

One of the conclusions of their study is that since without the pro region, there is no conversion between the I state and the native state (that is both states are stable), one of the states must be kinetically trapped. In general, protein folding is thermodynamically controlled [47,48], but in this case there is kinetic control.

3.4   The Hsp 70 chaperone/DnaJ system

Hartl [4] indicates that the Hsp 70s have a molecular mass of 70,000 ($M_r \sim$ 70K) and are a conserved class of ATPases. Their function includes the binding and release of hydrophobic segments of an unfolded polypeptide. The binding results in the stabilization of the polypeptide, and the release allows the peptide to progress along the folding pathway. The DnaJs or Hsp 40s modulate the action of the Hsp 70s. In this molecule there is a conserved J domain that interacts with its corresponding Hsp 70. This is a 70-amino-acid chain. Hartl [4] emphasizes that the J domain binds to different protein moieties in the Hsp 70/Hsp 40 system, and thereby not only protects the aggregation-prone polypeptides, but also facilitates specialized cellular functions [49–51]. Hartl [4] emphasizes that subtle differences in the J domains of the Hsp 40s are apparently critical in helping to mediate their respective Hsp 70 partners [48]. This author indicates that the central region in the J domain contains two zinc atoms and this along with a part of the C-terminal apparently binds to the unfolded poly-peptide. Furthermore, more than one Hsp 70 may interact with an unfolded polypeptide. Also, a critical part of the chaperone function is to present the misfolded aggregates of polypeptides to the cellular machinery for proteolytic degradation [52]. In order that one may obtain some further insights into *in vivo* protein folding, let us examine an example of *in vitro* folding.

3.5 Mechanism of β-hairpin formation and folding dynamics

Munoz et al. [53] have recently utilized a nanosecond laser temperature-jump method to analyze the folding kinetics of a β-hairpin that consists of 16 amino-acid residues. They monitored tryptophan fluorescence from a monomeric peptide G B1. These authors indicate that their analysis of β-hairpin folding in-cludes, in a broad sense, most of the basic physics of protein folding involving stabilization by hydrogen bonding and hydrophobic interactions, two-state be-havior, and a funnel-like, partially rugged energy landscape. A motivation for their studies was that although α-helix formation has been extensively studied in natural and synthetic peptides [54–56], the mechanism of β structure forma-tion has not been studied experimentally. Their analysis indicates that the for-mation of the hairpin occurs in 6 microseconds at room temperature. This is about 30 times faster than the rate of formation of the α-helix. Furthermore, their analysis of relaxation times of the unfolded and folded forms indicates that there is a single thermodynamic barrier that separates the unfolded state from the folded state.

4   Protein folding and fractals

Fractal kinetics have been reported in biochemical reactions such as the gating of ion channels [57, 58], enzyme reactions [59], and protein dynamics [60]. Fractals have also been utilized to analyze the binding kinetics of analytes in solution to receptors immobilized on biosensor surfaces [61–68]. It has been emphasized that the nonintegral dimensions of the Hill coefficient used to de-scribe the allosteric effects of proteins and enzymes are a direct consequence of fractal properties of proteins. Furthermore, Goetze and Brinkmann [69] sug-gest that a large number of systems in nature exhibit self-similar characteris-tics, ranging from molecular protein systems all the way up to astrophysical systems.

Presumably, folded and misfolded (or incorrectly folded) proteins would ex-hibit self-similarity at different levels of scale. Thus, the application of fractals to protein folding. Note that similarity does not imply identical. This self-simi-larity of the correctly folded form and misfolded forms one may characterize by different fractal dimensions, using suitable diagnostic or other methods. Perhaps, correct or incorrectly folded protein forms may be represented by different fractal dimensions or ranges of fractal dimensions. Basically, one may suggest that each stable state of the protein, either unfolded, correctly folded, misfolded, aggregated form, intermediates, etc. can be characterized by a fractal dimension or a range of fractal dimensions. As the protein accquires a higher degree of structure, it will be more heterogeneous in nature, and this protein structure would possess or exhibit a different fractal dimension. In other words, the folding process leads gradually to stable states that exhibit a lower entropy and a higher fractal dimension, whereas conversely, the unfolding process should lead to stable structures of lesser complexity that subsequently exhibit lower fractal dimension values along with higher entropic values as the protein chain unfolds. Also, during the folding process, correctly folded or native and active structures are formed along with aggregated forms. Since the stable and native protein form has more structure than the misfolded or aggregated forms, one may anticipate that the overall heterogeneity for the native and stable form is less than that of the aggregated form. This should then lead to higher values of the fractal dimension for the aggregated form compared to the native and active form. It would be of considerable diagnostic value if the ranges of fractal dimensions could be identified for the native and active form, as well as for the protein state when aggregated or misfolded forms are present. Note that this is over and above the inherent microheterogeneity that is present in the native and active protein state [70, 71].

Skinner [72] emphasizes that for chaos in biological systems only a few variables govern the spatial and temporal geometries of the system. An understanding of these fractal attractors or dimensions will significantly assist in the control of these systems. Cross and Hohenberg [73] have indicated that some systems that display spatiotemporal chaos are said to be 'large' since their desorption requires (more than one) a number of chaotic elements or 'attractors' distributed in space. The general impression about chaos theory is that one associates chaos with time as the independent variable, and fractals with space or a length dimension as the independent variable.

The fractal approach to characterize proteins is consistent with scaling laws and self-similar behavior that are apparently fundamental features of complex systems [74]. Proteins are appropriate objects or systems to analyze as complex systems. These authors indicate that many objects in nature are better described in the language of fractals than by Euclidean geometry. Ligand binding to the heme ion of Myoglobin (Mb) after flash disassociation has been analyzed [75]. These authors indicate that two states are clearly distinguished, the protein having the ligand bound or not. However, each of these states is comprised of a number of a hierarchy of substates that contribute to the transition kinetics by different rate constants or relaxation times. It is this distribution or heterogeneity of the different rate constants that leads to the fractal nature of the binding kinetics. On a similar basis one may suggest to utilize this hierarchical nature in the refolding of correct and misfolded forms of the protein to help characterize the different pathways observed and the stable forms prevalent in these sequences.

Buldyrev et al. [76] have analyzed the fractal landscapes and molecular evolution of the myosin heavy chain gene family. These authors indicate that nucleotide organization of DNA sequences have been analyzed considerably to provide a better understanding of local and global properties. These authors' analysis suggests that there is an increase in fractal complexity for these genes with evolution with vertebrate > invertebrate > yeasts. Thus, they validated their hypothesis that the fractal complexity of genes from higher animals is greater than that from lower animals, using single gene analysis. Other authors have indicated that there is a long-range power law correlation in certain DNA sequences that might extend over to 10^4 to 10^5 nucleotides [77–79]. Maddox [80] finds this of interest since this indicates the scale invariant nature of DNA sequences – a characteristic of fractal sequences. Once again, it is this nature of these types of systems that forms the basis and lends support to the suggestion that folded and misfolded forms obtained during the protein refolding process may be described by a fractal dimension, or by a range of fractal dimension values.

5   Conclusion

There is a critical need to minimize amyloid deposits or protein aggregation in humans. One needs to: a) identify compounds (presumably chaperones) in the body that will bind to these proteins (that aggregate) and move them along the correct direction towards the formation of the native and active protein(s), and; b) be able to activate these chaperones or other helping compounds in a synchronous fashion so that they can be of use as and when required. Note that these chaperones are multi-functional in the body. One of their functions is to help destroy the misfolded protein forms. This directly minimizes the aggregated forms, and prevents the accumulation of the deposits or entanglements that lead to the intractable diseases. Of course, one should also be able to turn these chaperones off. In other words, there should be a diagnostic or detection mechanism that signals that partially unfolded forms are above (or below) a certain threshold limit, and exhibit the tendency to aggregate and form deposits. Some evidence of this and the cellular machinery that exists is presented. More insight is required into *in vivo* folding mechanisms. Due to the scarcity of available information for *in vivo* protein folding/unfolding mechanisms (occurring under generally nonaqueous conditions) information available on *in vitro* protein (occurring under aqueous conditions and also under nonaqueous conditions) has been drawn upon to help better understand *in vivo* protein folding.

As more and more information is made available in the future about chaperones and their mechanisms of action, this and other aspects will get clearer. The folding aid or chaperone may also be presumably injected as and when required, though the means of delivery, dosage, and the side effects will have to be carefully studied. This is crucial due to the involvement of these chaperones in a mixture of reactions based on current knowledge available. As their role becomes clearer with the development of more knowledge about their action, it seems that they would appear to play an increasingly important role in helping maintain the activity and structure of a wide variety of proteins.

Furthermore, the body must possess a mechanism or a means by which it can identify correctly folded and misfolded protein forms in order that it may select and pick the right ones to repair. Some evidence for this is also presented. An additional means by which it may be able to do this would be by the use of immunoaffinity principles. In this procedure monoclonal antibodies directed against three-dimensional conformation-dependent epitopes select for correctly folded proteins [81]. If the proteins are not correctly folded, then the protein lacks the epitopes that are recognized by the conformation-dependent monoclonal antibodies. The unfolded forms must then be sent for further processing whereby they are either recycled to be refolded correctly, or destroyed if beyond repair.

The mechanisms for protein folding were presented. Basically, it is the series mechanism that leads to the native and active form. This is also consistent with the fractal nature of the system. It is the parallel or off-pathway steps that lead to aggregation and the misfolded forms. Means by which the correct (series) pathway is enhanced for example by the use of chaperones or folding aids are presented. Particular emphasis is placed on the cellular machinery that is in place, and how it functions to help proteins fold correctly and in the active form. More effort needs to be placed to help minimize these aggregates in both *in vivo* and *in vitro* studies. Considering that the chaperones will be involved in quite a few reactions, it would be worthwhile analyzing the dynamics of such reactions. The analysis of Robinson and Lauffenberger [37] is a step in this direction.

As mentioned earlier, the stakes are very high in increasing the correctly folded protein forms since this has direct biomedical implications in finding cures and strategies for the treatment of stubborn diseases. It may be recommended that not only should current treatment procedures be fully explored, but novel and unconventional treatment procedures should also be utilized to minimize amyloid deposits in human tissue. Some information with regard to this is presumably available within industrial sources, who understandably, are not too eager to share this information. There is indeed a large and ever expanding market for these block-buster drugs. This review hopefully brings this into focus, and indicates the need for further analysis in this direction along with the sharing of the information obtained by publication in the open literature, whereever possible.

6   Limitations and pitfalls of the models/approach

The basic mechanism for *in vivo* and *in vitro* protein folding/unfolding occurs via a series-parallel mechanism. It is the series mechanism that leads to the correct and active folded form. One needs to minimize the parallel steps that lead to the aggregate forms that not only minimize the yield of the active form (*in vitro)* but also lead to intractable diseases (*in vivo*). The mechanisms suggested are an initial attempt to understand protein folding mechanisms *in vivo*, and the influence of the medium on the protein folding/unfolding reaction. No feed-back steps or mechanisms were presented, since no information on this aspect is available. This is a limitation in the model and in the analysis presented. Surely, the human body does have feed-back mechanisms that help correct the formation of active proteins (especially the critical ones), and these come into play especially when the concentration of these critical proteins falls below a threshold level. Hopefully, as more knowledge is available in the open literature one will begin to understand *in vivo* protein folding reactions in different media a little better. Some of this information may be available in industrial sources, but considering the importance of drug development and the economic benefit in these critical areas, the reluctance of these sources to part with this information is understandable. Hopefully, the university sector will be able to fill this important and urgent need.

References

1   Ellis RJ (1990) Molecular chaperones: the plant connection. *Science* 250: 948–954

2   Rothman JE (1989) Protein folding and related processes in cells. *Cell* 59: 591–601

3   Gething MJ, Sambrook J (1992) Transport and assembly processes in the endoplasmic reticulum. *Nature* 355: 33–45

4   Hartl FU (1996) Molecular chaperones in cellular folding. *Nature* 381: 571–580

5   Selkoe DJ (1994) Alzheimer's disease beyond 1994: the path to therapeutics. *Neurobiology of Ageing* 15: 131–136

6   Jen LS, Hart AJ, Jen A et al (1998) Alzheimer's peptide kills cells of retina *in vivo*. *Nature* 392: 140–141

7   Vogel G (1998) Tau protein mutations confirmed as neuron killers. *Science* 280: 1524–1525.

8   Borman S (1998) Prion research accelerates. *Chem Eng News* February 9, 22–29

9   Mitraki A, King J (1989) Protein folding intermediates and inclusion body formation. *Bio/ Technology* 7: 690–696

10   Lazardis T, Karplus M (1997) "New view" of protein folding reconciled with the old through multiple unfolding simulations. *Science* 278: 1928–1931

11   Chan HS (1998) Matching speed and locality. *Nature* 392: 761–763

12   Hlodan R and Hartl FU (1994) How the protein folds in the cell. In: RH Pain (ed).: *Mechanisms of protein folding.* IRL Press, Oxford, England, 194–228

13   Kane JF, Hartley DL (1998) Formation of recombinant protein inclusion bodies in *Escherichia coli*. *Trends Biotechnol* 6: 95–101

14   Hartl DL, Taubes CH (1996) Compensatory nearly neural mutations. Selection without adaptation. *J Theor Biol* 182 (3) 303–309

15   Pfeifer P, Obert M (1989) Fractals: basic concepts and terminology. In: D Avnir (ed): *The fractal approach to heterogeneous chemistry: surfaces, colloids, polymers.* J. Wiley & Sons, New York, 11–43

16   Schmid FXA (1993) Kinetics of unfolding and refolding of single-domain proteins. *Rev Biophys Biomol Struct* 22: 123–143

17 Jaenicke R (1991) Protein folding: local structures, domains, subunits, and assemblies. *Biochemistry* 30: 3147–3161

18 Go N (1983) Randomness of the process of protein folding. *Int J Peptide Res* 22: 622–632

19 Mendoza JA, Rogers E, Lorimer GH, Horowitz PM (1991) Chaperonins facilitate the *in vitro* folding of monomeric rhodanese. *J Biol Chem* 266: 13044–13049

20 Pitsyn OB, Reva BA, Finkelstein AV (1989) Prediction of protein secondary structure based on physical theory. *Highlights Mod Biol* 1: 1–18

21 Jaenicke R (1987) Folding and association of proteins. *Progr Biophys Mol Biol* 49: 117–237

22 Goldenberg DP, Creighton TE (1985) Energetics of protein structure and folding. *Biopolymers* 24: 167–182

23 Kotlarski N, O'Neill BK, Francis GL, Middleberg AP (1997) Design analysis for refolding monomeric protein. *A I Ch E J* 43(8): 2123–2132

24 Ghag SK, Brems DN, Hassel TC, Yeh WK (1985) Solubility of different folding conformers of bovine growth hormone. *Biotechnol Appl Biochem* 24: 167–182

25 Haase-Pettingel C, King J (1988) Formation of aggregates from a thermolabile *in vivo* folding intermediate in P22 tailspike maturation. *J Biol Chem* 237: 1839–1845

26 Sadana A (1991) *Biocatalysis: Fundamentals of enzyme deactivation kinetics.* Prentice Hall, Englewoods Cliffs, NJ, USA

27 Sadana A, Malhotra A (1987) Effect of activation energy microheterogeneity on first-order enzyme deactivation. *Biotechnol Bioeng* 30: 108–117.

28 Batas B, Chaudhari JB (1996) Protein refolding at high concentration using size-exclusion chromatography. *Biotechnol Bioeng* 50:16–23

29 West SM, Chaudhari JB, Howell TA (1998) Improved protein refolding using hollow-fibre dialysis. *Biotechnol Bioeng* 57: 590–599

30 Hashimoto Y, Ono T, Goto M, Hatton TA (1998) Protein refolding by reversed micelles utilizing solid-liquid extraction technique. *Biotechnol Bioeng* 57: 620–623

31 Moutiez M, Guthapel R, Gueguen P, Quememeur E (1997) New formulae for folding catalysts make them multi-purpose enzymes. *Biotechnol Bioeng* 57: 590–599

32 Middleberg APJ (1996) The influence of protein refolding strategy on cost for competing reactions. *The Chemical Eng J* 61: 41–52

33 Cleland JL, Wang DIC (1991) Co-solvent effects on refolding and aggregation. In: ME Himmel, G. Georgiou (eds): *Biocatalyst design for stability and specificity.* ACS symposium series 516, American Chemical Society, Washington, DC, USA

34 Walshaw J, Goodfellow JM (1993) Distribution of solvent molecules around apolar side-chains in protein crystals. *J Mol Biol* 231:392–414.

35 Powers ME, Matsuura J, Brassel J et al (1993) Enhanced solubility of and peptides in nonpolar solvents through hydrophobic ion pairing. *Biopolymers 27:* 927–932

36 McMinn JH, Sowa MJ, Charnick SB, Paulitis ME (1993) The hydration of proteins in nearly anhydrous organic solvent suspensions. *Biopolmers* 33: 1213–1224

37 Robinson AS, Lauffenberger DA (1996) Model for ER chaperone dynamics and secretory protein dynamics. *A I Ch E J* 42(5): 1443–1453

38 Braakman S, Helenius J, Helenius A (1992) Role of ATP and disulfide bonds during protein folding in the endoplasmic reticulum. *Nature* 356: 260–262

39 Munro S, Pelham, HRB (1986) A C-terminal signal prevents secretion of luminal ER proteins. *Cell* 46: 291–300

40 Flynn GC, Chappel TG, Rothman JE (1989) Peptide binding and release by proteins implicated as catalysts in protein assembly. *Science* 245: 385–390

41 Pelham HRB (1986) Control of protein exit from ER. *Cell* 46: 1443–1453

42 Kozutsumi Y, Segal M, Normington K et al (1988) The presence of malfolded proteins in the endoplasmic reticulum signals the induction of glucose-regulated proteins. *Nature* 332: 462–464

43 Sharma S, Rodgers L, Brandsma J et al (1985) SV40 T antigen and the exocitic pathway. *EMBO* 4: 1479–1485

44 Baker D, Sohl J, Aagard DA (1992) A protein-folding reaction under kinetic control. *Nature* 356: 263–265

45 Silen JL, Frank D, Fujishige A et al (1989) Analysis of prepro-α-lytic protease expression in *Escherichia coli* reveals that the pro region is required for activity. *J Bact* 171: 1320–1325

46 Silen JL, Agard DA (1989) The α-lytic protease pro-region does not require a physical linkage to activate the protease domain *in vivo*. *Nature* 341: 462–464

47 Kim P, Baldwin RL (1990) Intermediates in the folding reaction of small proteins. *Biochemistry* 59: 631–660

48 Dill KA, Alonoso DOV, Hutchinson K (1989) Thermal stabilities of globular proteins. *Biochemistry* 29: 133–155

49 Sanders SL, Whitfield KM, Vogel JP et al (1992) SEc61p and BiP directly facilitate polypeptide translocation. *Cell* 69: 353–365

50 Schlenstadt G, Harris S, Risse B et al (1995) A yeast DnaJ homologue, Sc1p, can function in the endoplasmic reticulum with BiP/Kar2p via a conserved domain that specifies interactions with Hsp70s. *J Cell Biol* 129: 979–988

51 Ungewickell E, Ungewickell H, Holstein SEH et al (1995) Role of auxilin in uncoating clathrin-coated vesicles. *Nature* 378: 632–635

52 Hayes SA, Dice FJ (1996) Role of molecular chaperones in protein degradation. *J Cell Biol* 132: 255–258

53 Munoz V, Thompson PA, Hofrichter J, Eaton WA (1997) Folding dynamics and mechanism of β-hairpin formation. *Nature* 390: 196–199

54 Zimm B, Doty P, Iso K (1959) Determination of the parameters for helix formation in poly-γ-benzyl-L-glutamate. *Proc Natl Acad Sci* 45: 1601–1607

55 Gruenewald B, Nicola CU (1979) Kinetics of the helix-coil transition of a polypeptide with non-ionic side groups derived from ultrasonic relaxation measurements. *Biophys Chem* 9: 137–147

56 Chakrabartty A, Baldwin R (1955) α-helix stability. *Adv Prot Chem* 46: 141–176

57 Liebovitch LS, Sullivan JM (1987) Fractal model of a voltage-dependent potassium channel from cultured mouse hippocampal neurons. *Biophys J* 52: 979–988

58 Liebovitch LS, Fischbarg J, Koniarek JP et al (1987) Fractal model of ion-channel kinetics. *Math Biosci* 84: 37–68

59 Li H, Chen S, Zhao H (1990) Fractal mechansims for the allosteric effects of proteins and enzymes. *Biophys J* 58: 1313–1320

60 Dewey TG, Bann JG (1992) Protein dynamics and 1/f noise. *Biophys J* 63: 594–598

61 Sadana A (1995) Antigen-antibody binding kinetics for biosensors: the fractal dimension and the binding rate coefficient. *Biotechnol Prog* 11: 50–57

62 Sadana A, Sii D (1992) The binding of antigen by immobilized antibody: influence of a variable rate coefficient on external diffusion limitations. *J Colloid Interface Sci* 151: 166–177

63 Sadana A, Madagula A (1993) Binding kinetics of antigen by immobilized antibody or of antibody by immobilized antigen: influence of lateral interactions and variable rate coefficients. *Biotechnol Prog* 10: 291–298

64 Sadana A, Madagula A (1994) A fractal analysis of external diffusion-limited first-order kinetics for the binding of antigen by immobilized antibody. *Biosens Bioelectron* 9:45–55

65 Sadana A, Beelaram A (1994) Antigen-antibody binding kinetics for biosensors: the fractal dimension and the binding rate coefficient. *Biotechnol Prog* 10: 291–298

66 Sadana A, Beelaram A (1995) Antigen-antibody diffusion-limited binding kinetics of biosensors: a fractal analysis. *Biosens Bioelectron* 10: 301–316

67 Sadana A, Sutaria M (1997) Influence of diffusion to fractal surfaces on the binding kinetics for antibody-antigen, analyte-receptor, and analyte-receptorless (protein) systems. *Biophys Chem* 65: 29–44

68 Sadana A (1998) An analysis of analyte-receptor binding kinetics for biosensor applications: influence of the fractal dimension on the binding rate coefficient. *J Colloid Interface Sci* 198: 164–175

69 Goetze T, Brinkmann J (1985) Self similarity of protein surfaces. *Biophys J* 61: 109–115

70 Caan JR (1982) Theory of sedimentation for antigen-antibody reactions: effect of antibody heterogeneity on the shape of the pattern. *Immunol* 19: 505–512

71 Caan JR, Fink NH (1983) Effect of microheterogeneity on the sedimentation behavior of self-associating proteins. *Biophys Chem* 17: 29–36

72 Skinner JE (1994) Low dimension chaos in biological systems. *Bio/Technology* 12: 596 -600

73 Cross MC, Hohenberg PC (1994) Spatiotemporal chaos. *Science* 263: 1569–1573

74 Glockle WG, Nonenmacher TF (1995) A fractional calculus approach to self-similar protein dynamics. *Biophys J* 68: 46–51

75 Alberding N, Austin RH, Chan SS et al (1992) Dynamics of carbon monoxide binding to protoheme. *J Chem Phys* 65: 4701–4711

76 Buldyrev SV, Goldberger AL, Havlin S et al (1993) Fractal landscapes and molecular evolution : modeling the myosin heavy chain gene family. *Biophys J* 65: 2673–2679

77 Peng CK, Buldyrev SV, Goldberger AL et al (1992) Fractal landscape analysis of DNA walks. *Phys Rev Lett* 47: 3730–3736

78 Voss R (1992) Evolution of long-range fractal and 1/f noise in DNA base sequences. *Phys Rev Lett* 68: 3805–3812

79 Munson PJ, Taylor RC, Michaels GS (1992) DNA correlations. *Nature (Lond.)* 360: 636–640

80 Maddox J (1992) Long-range correlations. *Nature (Lond.)* 358: 103–106

81 Wells PA, Biedermann B, Garlick RL et al (1994) Large-scale immunoaffinity purifications of recombinant soluble antigen CD4 from Escherichia coli cells. *Biotechnol Appl Biochem* 18: 341–346

Guide to Solutions

Guide to Enzymes

Guide to Protocols

Troubleshooting Guide

Index